CFZ YEARBOOK 2002

edited by
Jonathan Downes and Richard Freeman

Typeset by Jonathan Downes,
Cover and Layout by SPiderKaT for CFZ Communications
Using Microsoft Word 2000, Microsoft , Publisher 2000, Adobe Photoshop CS.

Photographs © 2007 CFZ except where noted

First published in Great Britain by CFZ Press

**CFZ Press
Myrtle Cottage
Woolsery
Bideford
North Devon
EX39 5QR**

© CFZ MMVIII

All rights reserved. Without limiting the rights under copyright reserved above, no part of this publication may be reproduced, stored in or introduced into a retrieval system, or transmitted, in any form of by any means (electronic, mechanical, photocopying, recording or otherwise), without the prior written permission of both the copyright owners and the publishers of this book.

ISBN: 978-1-905723-30-0

This book is dedicated to the memory of

Roly and Lee Holloway

(for many years the best neighbours the CFZ could possibly have had)

Contents

7   Introduction to the 2008 edition
9   Introduction
11 2001 – a year in the life of the Centre for Fortean Zoology
13 *Behind the Legend of Boggy Creek* by Neil Arnold
29 *Morgawr the Sea Dragon – an overview* by Jon Downes
49 *The Dobhar Chu – Ireland`s currently enigmatic and sometimes deadly unknown otter* by Gary Cunningham
85 *The Complete Mystery Reptiles* by Jonathan Downes
103 *Hybridisation in the family felidae* by Chris Moiser
109 *British Big Fish Records* by Neil Arnold
124 *Fortean Zoological Aspects of the Biggles books of Captain W.E.Johns* by Graham Inglis with additional notes by Richard Freeman
149 *Cassowarys – The meanest, coolest birds alive* by Darren Naish

Introduction
To the 2008 Edition

I *am* enjoying overseeing this reissue programme of the CFZ Yearbooks, although - if I am brutally honest - like even the most enjoyable tasks, the effort becomes more enjoyable as the end looms in sight, and with the completion of this present volume, there is only one -1997 - left to do, and that should be out within the next few weeks.

However, it has been a generally edifying experience to look back over the fourteen years since I first came up with the idea of the CFZ Yearbook. Occasionally it has been a mildly embarrassing one as well, as with the 1999 yearbook when we found a shed load of spelling mistakes and other typos with which we had to deal.

But editing (or, I suppose, to be strictly accurate: re-editing) this yearbook was a highly pleasurable experience. Back in 2002, as the CFZ was beginning to actually make its mark on the world's stage, we were finally beginning to ride the learning curve, and we not only knew what we wanted to do, but we knew how to do it.

The history of the CFZ Yearbooks to date can really be divided into two, and it was this volume which kicked off the second stage. I think that it was our annual conference; The Weird Weekend that did it. This was our big attempt at building a proper cryptozoological community, and on the whole we were (and have been in the years since) successful

The contributors to this volume are all, with one exception, people who have appeared at the Weird Weekend over the years, and the one exception (Neil Arnold) is pencilled in for the 2008 and 2009 events.

With this volume, and with the attendant Weird Weekend, the CFZ finally began to grow up. After a long and painful adolescence it was time for us to grow up and take our place with the big boys in the real world.

Within a couple of years we were unrecognisable. We were mounting expeditions all over the world, publishing an impressive range of books and magazines, had an interest in a local zoo, and were building our own museum.

But looking back, it all started with this yearbook!

Enjoy…

Jon Downes,
Centre for Fortean Zoology,
North Devon
May 2008

Introduction

Dear Friends,

Welcome to our sixth yearbook. For those of you not aware of the history of this series, we have been publishing *Animals & Men* (our quarterly journal) since 1994. However, in 1995 we realised that much of the material submitted to us was far too long for a quarterly 48pp magazine so we instituted the Yearbook series in order to be able to publish longer research papers and expedition reports. In many ways we feel that this volume is the best yet. It includes, what we believe, to be the first ever in-depth investigation into the *Dobhar Chu* – the semi legendary Master Otter of Ireland. It also contains Neil Arnold's remarkable investigation into the genesis of the Fouke Monster legends and much more. We are very proud to be able to present such a dazzling array of fortean and cryptozoological talent and hope that this volume will be a benchmark by which future volumes can be measured.

Slainte Mhor

2001 - a year in the life of the Centre for Fortean Zoology

After the disastrous events of 2000, this was the year that we finally managed to get back on track. The end of 2000 saw the Centre for Fortean Zoology facing bankruptcy and after a year of bereavements and disasters we were tired and unsure even whether we were going to be able to carry on. Then help came in the shape of a loan from our dear friend Joyce Howarth, and we began to pull ourselves up by our proverbial bootstraps.

In May we promoted our second annual convention – the Weird Weekend (we had no idea that Louis Theroux had already coined the title honest) - and throughout the year we became involved in a succession of projects ranging from projected expeditions to collaborations with the English Folkdance Project into the zoomorphic archetypes used in ritual dancing.

I suffered ill health throughout most of the year but was well on the way to recovery as we went to print with this volume. In the summer, our Hon Consulting Editor, Bernard Heuvelmans, often known as *The Father of Cryptozoology* died. After much soul searching we invited Colonel John Blashford-Snell to be his replacement. He has justly been described as Britain's real-life Indiana Jones and is a fitting symbol for an organisation that we hope will be blazing a cryptozoological trail well into the 21st Century and hopefully beyond.

The future is bright.....

Behind the legend of Boggy Creek

by Neil Arnold

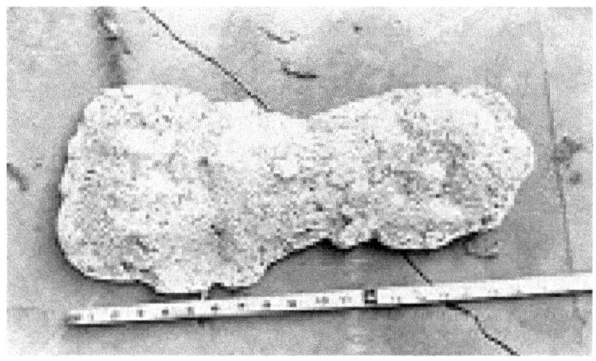

Photo of footprint found 27th November 1973.

- Fouke, Arkansas. Population 634 (265 families). Positioned 33.26 degrees north of the Equator and 93.88 degrees west of the prime meridian.
- Land area in Fouke is 2.713 kilometers. The town is 1058 miles from Washington and a distance of 136 miles to the Arkansas state capital.
- Springs, quagmires and swamps bordered by Louisiana, Mississippi, Missouri and Oaklahoma. Heavily populated by giant hogs, beavers, bullfrogs and up until 1991 one unidentified ape-like creature.
- Neil Arnold. 1982. Kent, England. 7 years-old at the time watching BBC 2 with parents. Terrified and yet fascinated by the docu-drama on television which was to become one of the most sought after cult films of all time. "The Legend Of Boggy Creek".
- Charles B. Pierce. Film director. With a budget of $160,000 combined fact and fiction to

make the film, set in the Fouke creeks around 1971.
- Charles Williams. 2001. Kent, England. Revives the legend of the film and the mystery by buying the rights to the Boggy Creek name and bringing the film and the facts to the masses
- The legend is…a hairy Bigfoot terrorised the town and surrounding areas in the 1970s, seemed immune to gunfire and used the heavy foliage and woodlands along the creeks to conceal its presence.
-

The truth is as follows….and it is far more startling than the fiction.

As a film, "The Legend Of Boggy Creek" remains one of, if not the most, frightening ever made. Constructed as some sort of rough, home made, low-budget spook documentary it embedded itself in thousands of people's brains.

Many of them now just wondering where that film is, especially as similar films such as "The Blair Witch Project" are raking it in at the box office. The film is just as elusive as the sasquatch itself, but the facts remain hidden, obscured by the dramatics and yet they could well be the most startling details of a legacy that continued until 1991.

Smokey Crabtree is the main character around which the Boggy Creek phenomenon revolves. He is a man who has spent all of his life carving a living from the land around him. He is also an honest man who wishes away the past with regards to the media presence bestowed upon the local community whilst the film was made.

Smokey has told his main story in his book, *"Smokey And The Fouke Monster"*, first published in 1974 and recently reprinted. The aim of this article is to present the 'monster' for what it really was…a real beast, which, only now - some forty years later - has come to light as a remarkable case, ignored by science and unknown to cryptozoologists.

Creature sighted here 27th May 1971

Smokey Crabtree may well appear as a typical backwoods grafter, with a country twang in his vocal chords and a career of strict labour which took him from Canada to Saudi Arabia, in order to provide for his family. Smokey has always respected man and nature, living off the latter and working hard for the former to gain respect as an honest working man. No-one ever crossed Crabtree and got away with it, and no-one ever knew the bottoms of Arkansas like him…except for the hairy stranger.

After many years of broken dreams and false ambitions, Smokey and his family (wife June, three sons and a daughter) settled in Jonesville in 1962. A business in Fouke had been sold and so 480 acres of land at Days Creek would welcome them. The family built their own house and an 80 acre lake and Smokey was often called away for pipelining employment which generally fed his family.

During the year of 1965 Smokey's second son James Lynn had an encounter that would change the Crabtree's lives forever. Their seclusion and idyllic residence would never be the same again.

Whilst out squirrel-hunting in the bog lands and underbrush behind the lake one morning, James was alerted to the sound of horses running down an old logging road. The horses were some seventy-five yards away, but he clearly heard them rush into the lake.

James knew that horses roamed the woodlands and simply thought the pesky flies had irritated them. However, the howl of Lynn's squirrel dog near the horses alarmed him, and thinking the dog was hurt he rushed to the scene. Upon arriving he was stopped dead in his tracks by a deep grunting. The guttural growl was unlike anything he had heard before and with only his .20 gauge shotgun by his side he felt vulnerable to this presence. Then, moving into an opening he saw, only thirty steps away, a huge hairy animal of some kind with his back towards Lynn. The 'thing' was gazing at the horses and appeared to have run them into the lake before stopping at the edge.

James Lynn was amazed by the monstrosity, which stood around seven-feet tall, was covered in reddish-brown hair roughly four inches long and stood upright. The arms were long as they hung at the side and the creature appeared disgruntled and agitated before turning away from the horses. After a few steps the beast noticed James. It reached into the air with its head as if to sniff the breeze and then it slowly headed towards the frightened fourteen-year old. It was at this point that the remarkable happened as James aimed his gun at the approaching man-beast and fired at its face. The creature seemed immune to the bullets, taking a couple more shots before James sprinted away in terror.

When James Lynn reached home he was in a state of extreme panic. Smokey got the story out of him and found such a yarn difficult to absorb. Strangely enough though, Smokey had heard that a neighbour some three miles away had also caught sight of such a hairy creature as it peered through the window of their home.

Whatever had spooked Lynn and the horses seemed to have visited the lake not for the sake of frightening any man or animal, but merely to quench its thirst. Smokey and a group of locals took to the area at night believing that the creature may well return for a drink, and despite a few weird atmospheres nothing materialised.

Within days Texarkana reporters were on the scene, but Smokey hushed them away saying that such a thing would have to be seen to be believed. Such harsh words rattled his son and also caused a shock confession from Smokey's uncle James who claimed that he saw the same thing just a few months before near the river. The creature was apparently cleaning its feet and then noticed James before running off into the thick woodlands. James also described the animal as man-like, covered in hair and large.

A week or so after James' incident, a cousin of Smokey also spoke of a peculiar sighting. Fred Crabtree had almost stumbled upon the shaggy thing, which stood between seven and eight feet tall, but once again the beast disappeared into heavy foliage.

The Jonesville Monster had, all of a sudden, been born, although most folk believed the creature to be a bear or errant gorilla. Within the month though another boy, Kenneth Dyas, aged fourteen, also shot at a similar looking creature whilst deer hunting in the same woods, and ran like the wind for home.

The next sighting of the creature involved a female deer hunter from the Fouke area. Whilst alone on her deer stand and with the rest of the hunters on a widespread search, the creature seemed almost accidentally to come by, causing the witness to cry for help and run to her vehicle for safety. It was an encounter a year or so after James Lynn's incredible sighting and it was not an incident that attracted too much media attention, however the male hunters tracked the hairy beast with their dogs in an area only one mile from the reports of James and Fred Crabtree.

Despite the rising fear in the community, there was never any evidence to suggest that the hairy abomination was out to cause harm, even in its apparent primal state. Weeks after the deer hunter encounter a local school bus driver on a rural, quiet route saw the bipedal beast run across the road ahead of his vehicle early one morning. All the sightings appeared to be in the same sort of area.

It was then that Smokey began to truly believe in the existence of the swamp creature. On several occasions in the dead of night Crabtree heard a large creature outside the house which would omit a terrible scream, sending the dogs into panic and causing them to flee the backyard for days. The cattle would also act peculiar; giving Smokey reason to believe the beast was back and certainly nearby as the animals stampeded through the woods alarmed by some unseen presence. On a number of occasions up until 1971, the beast would linger in the Crabtree's yard, its awful wail soaring across the woodlands. Yet not once did the family report the presence.

In the same year 25-year-old Bobby Ford reported to a Fouke police officer that he'd heard screams of terror from womenfolk after a large, hairy creature poked its head through the window of their house. Ford allegedly investigated the sighting, which took place in the May and was supposedly attacked by the monster as he returned to the house. According to a report in the ARKANSAS GAZETTE dated May 4th 1971 (Tuesday), *"...he was attacked by a large, hairy creature at his home and was treated at a hospital here early Sunday for scratches and shock"*.

The newspaper also goes on to mention that, *"...Ford said something kicked in the back door and they again saw the creature behind the house. As he was returning to his house after chasing the creature, he said the animal knocked him down..."*.

Smokey Crabtree immediately dismissed such a story although did believe that the creature had moved to Fouke, but not to become so savage all of a sudden. The Ford case remains dubious but there were further sightings at Shreveport, Louisiana and Texarkana, with one report mentioning another road encounter where three residents saw the creature as it crossed Highway 71. The animal was described as being seven feet tall, and faster than a man.

The Jonesville Monster became the Fouke Monster in a matter of months, with the locals keen on getting a piece of the creature's flesh. It was around this time also that the legend of Boggy Creek was born. A legend that never existed, but which obviously appealed to movie director Charles B. Pierce. It was this man who barged into Crabtree's life looking for the monster and literally stole the mystery, eventually blowing it out of proportion and turning the local community into a theme park for monster hunters and tourists. But back to the sightings...

June 1971 brought with it one of the most remarkable pieces of evidence with regards to the Fouke Monster. Smokey was called, by his brother-in-law Willie Smith, one morning in June. Willie asked Smokey to come along to his place and to bring a big rifle. Excitedly Smokey rushed to the scene and was immediately dragged to Smith's soybean field where he and Mr. Kennedy had found some peculiar tracks. The field was around three inches high with beans and freshly ploughed. The land according to Smokey was, "...sandy and extra clean".

Smith and Kennedy had discovered the tracks after a rain shower and there were hundreds of them. Straightaway, Smokey was bemused by the unusual tracks like which he had never seen before. The tracks were thirteen-and-a-half inches long, four-and-a-half inches wide at the ball and three-and-a-quarter inches wide at the heel. The instep was very high, and where the sand was extra shallow the heel part of the track was not connected to the front part. Smokey went on to describe the tracks:

"Where the sand was one-half inch deep or so there was a narrow outer part of the foot touching down, connecting the heel to the front part with a one-inch wide or so strip. Where the dirt was soft there was a full track and very plain ones. They were so plain you could see the imprint of the lines in the bottom of its bare feet. There were only three toes and there wasn't all that much difference between the size of the first toe and the third toe...his foot was designed for three toes only".

According to Smokey also, the creature did not seem to know what it was doing in the field. It came out of the woods and walked right out into the field. The stride measured 57 inches from heel to toe at first and then became inconsistent, at times being as low as 26 inches. Smokey also believed the 'monster' was female due to the high instep and narrow nature of the foot. Smokey estimated the creature to weigh around three hundred pounds.

The tracks veered left in the field, occasionally stopped and eventually came around in a circle before they returned to the exact same spot in the woods. Such tracks, and the animal's hollering were the closest Smokey ever got to the creature.

The tracks created a lot of publicity and monster-hunters from around the United States loitered in the Jonesville and Fouke woodlands, pestering Smokey and his family, whilst the insistent Charles B. Pierce made his intentions clear that he was going to make a movie about the mystery, and Smokey was to be the main informant and middle man to drag in the local folk to speak about their bizarre encounters.

Originally the idea appealed to Smokey, a wage was agreed so that Pierce and co. could use the area to interview witnesses, and to employ Crabtree as a sort of wildlife guide for nightly visits into the swamps. Smokey certainly wasn't willing to lie for anyone or put the reputations of his neighbours on the line and entangle them in some fictional monster horror story. As far as he knew, Pierce and film crew were out to make a factual, honest broadcast.

It is from here that the Boggy Creek story becomes a major quibble fuelled by lawsuits, broken and fictional contracts, deceit, mistrust and an eventual breakdown of friendships within the community. Whether the monster still roamed the creeks became second-place to a commotion that Smokey still has failed to appease some thirty years later. The legend of Boggy Creek as a film, despite its suspense is overblown and mostly fictional. There are various situations within the plot, which never took place at all, and there are various inclusions that were not consented to by local folk. Smokey completely withdrew himself and his family from the proceedings whilst Charles B. Pierce made mega-bucks from a patchy fantasy.

"*The Legend Of Boggy Creek*" premiered August 18th 1972 in Texarkana, bereft of truth, and whilst the more interesting monster sightings were still happening. The following year a non-believer of the creature, 67-year old Orville Scroggins saw the animal in his field four miles outside Fouke. Orville was with his son and grandson at 7:00am during November delivering a calf from its mother when they heard a few cows becoming disturbed. Orville looked up and saw a hairy creature one-hundred yards away walking east. He described the creature as, "... *four feet tall, stood up-right and had long, pitch black hair*".

Prints were found in the area spanning five-and-a-half inches in width and were forty inches apart.

Many believed the Scroggins sighting because he had become perceived as the main sceptic towards the creature's existence. However, the details of the sighting with regards to the height of the animal differ greatly from the other reports. As recently as 1997, there was belief in a family of hairy bipeds inhabiting Fouke although more rational folk would argue that the

1973 sighting was a hoax, either perpetrated by a local, or by Pierce himself, who allegedly roamed the area planting false prints.

According to available evidence of the creature, the sightings seemed to come to an abrupt halt after 1974 when a seven-foot tall "gorilla-like" creature was spotted at Holly Springs, a few miles out of town. The truth is there were a number of reports from Fouke in the early 1980s. However, Smokey Crabtree never once heard of a sighting within the area of Boggy Creek, but Charles B. Pierce's original film title of *"Tracking The Fouke Monster"* obviously didn't have the same ring to it.

Various articles in the 1990s, and through to the millennium, have dismissed the legend as a whole, blaming the film for local hysteria and even condemning Smokey for the unwanted attention he bestowed upon the community. Smokey is attempting to straighten out the very rough edges but no-one is willing to take the case to the courts to fight Pierce and the money men. The actual facts of the later years concerning the monster are unknown...until now.

The legend of Boggy Creek quite literally died in the late '80s, or possibly early '90s, despite the claims of some that they still hear the monster's terrible howl. The monster of Fouke was a one-off and its corpse has been found intact except for the skull and is being studied by scientists and kept in a secret location. I obtained this information from Kent man Charles Williams, the man who has bought the rights to the "...*Boggy Creek*" film and the web sites, thus allowing the true story of Smokey Crabtree to reach a greater audience.

The film, since its release, has reaped £100 million and a remake is in the pipeline. Through Charles the book *"Smokey & The Fouke Monster"* tells the story of the extraordinary man's perseverance through thick and thin and sharing his backyard with the hairy monster. Charles tracked down the owner of the docu-drama to Florida, which enabled the movie to reach DVD and the truth of the story come to light.

In 1991, the remains of an unknown bipedal creature were found by a group of children playing in a bean field just a few miles outside of Fouke. Smokey was called to take a look at the remains and took the skeleton away. The skeleton measured a staggering seven-feet in length without the head which may well have sunk into the swampy area or been taken away by wolves as the carcass had been badly eaten. Analysis revealed that the creature had died of pneumonia.

I have personally seen photographs of the skeleton, and video footage, and it is a skeleton that resembles no other human or primate. It has a spine length of 51 inches, and long flat feet with hooked toes completely unlike the typical sasquatch prints. The DNA structure of the remains matches no corresponding DNA structure of known man or beast.

So, what is the identification of the real Fouke Monster? Where did it come from? Well, whilst many bigfoot sightings attempt to spotlight missing links, Neanderthal connections or men in fur suits, the truth behind the Boggy Creek monster is far more bizarre and intriguing. During the early 1960s a train carrying a travelling circus crashed ten miles outside of Fouke. Allegedly, the circus carried cages of freaks and curiosities and it is believed that such a hybrid or odd-body escaped from the wreckage, eventually inhabiting the vast woodlands and misty creeks. Rumour has it that the local Sheriff's department knew of the crash, but stayed quiet with regards to the escaped contents of the freight train.

The beast of Fouke was a peculiar manifestation that did not really resemble the sasquatch form despite reports of the shaggy hair. The fact that bigfoot corpses are never found seems to suggest that such creatures may well crawl away to die before their flesh is eaten by natural predators. The Fouke Monster died of a natural death on swampy, but accessible terrain…and it was found.

Local newspaper reporters from Miller County insist the legend continues, claiming that forty reports were taken towards the end of the 1990s, but none of them know the truth. The area and its monster-hunters exist on the mystery, many claiming that the creature existed back in the '40s but no evidence has come forward to suggest such a thing. The fact that stickers, t-shirts, caps and badges of the creek monster still circulate in local stores shows how dedicated the people are, despite the opposing views of those unfortunate townsfolk who have had enough of the creature and the tourists who stampede through the swamps and hide out in the backyards of unsuspecting locals, hoping to catch a glimpse of the elusive thing.

Imagine the response of such downbeats if they were told the monster of Fouke is dead and had been for some ten years. Only a 1996 documentary called "The Hunt For Bigfoot" has officially exposed the truth whilst the rest of us still believe that Bigfoot inhabits Arkansas, and some of us never have. Neither view points were correct.
Fifty percent footage of the "..Boggy Creek" film was shot after Smokey Crabtree was fired from the project. Pierce and company moved out of the Fouke area yet continued to shoot footage which allegedly took place in the area.

Charles B. Pierce was broke when he made the film…now he isn't!

Smokey Crabtree still lives in the creeks and wilderness of Arkansas. Charles Williams has copies of the "*Legend Of Boggy Creek*" for sale on DVD and video. They can be obtained at www.thelegendofboggycreek.com and www.moviesunlimited.co.uk

Neil Arnold still recalls the childhood nightmares caused by the hairy man-beast, which allegedly terrorised the Fouke community and cowered under its terrifying screams. Many other movie buffs have had similar nights of unrest. And in the famous words of dear old Smokey, "*Well, I have not slept much since the picture was released myself*"!

"HE ALWAYS WALKS THE CREEKS…"

Sightings that have taken place after 1991 have been classified as hoax. However, it would not be a surprise for genuine bigfoot phenomena to occur in the heavy swamplands where only beaver, bullfrog and snapping turtle dare travel. Large unknown footprints were allegedly found from the late '70s right up until the present day, and in-between there are other reports

CFZ Yearbook 2002

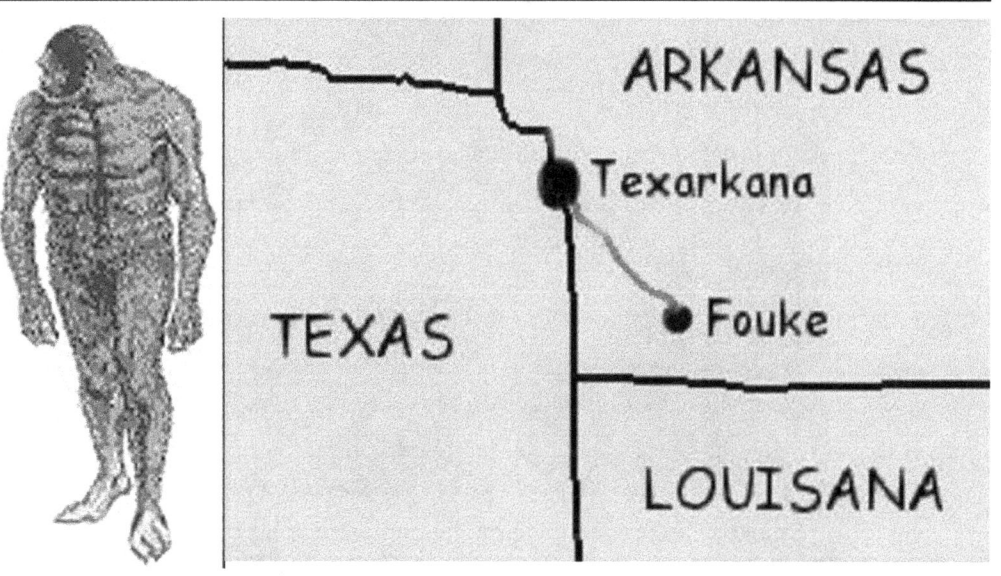

Artist's impression of the Fouke monster, map , and cover of the video which you can buy :

www.
thelegendofboggy-
creek.com

Smokey Crabtree hunting on the creeks near his home

of a hairy creature, similar in description to the known Fouke creature, which seem to fill in the void that many researchers have leapt across whilst eye-witness reports were quiet in the Arkansas region.

The Fouke beast may well have been the only one of its kind, a lonesome, frustrated abomination or freak of nature, and certainly a creature treated harshly by the hunters, as it would have been in its cankered captivity. Smokey Crabtree never once believed that the hairy inhabitant of Fouke was any type of 'monster'. The reports of mutilated chickens, disembowelled dogs and slaughtered pigs were mere sensationalism surrounding a creature that more than likely fed from the Creek beds on berries, foliage and general wildlife within the woodlands.

The story of the Fouke Monster is more akin to "The Elephant Man" than some cryptozoological riddle. There are many tales of freak carnivals and bizarre exhibitions, which travelled cross-country, like some obscure peep show, parading the more normal women with beards and children with flippers, but there was always the possibility that some diseased or deformed mutation would have been taken on board to pull in the punters. Remember "*King Kong*" and "*Mighty Joe Young*"? Mere Hollywood monsters or real, caged beasts?

In an English chat show called "Esther", famous actor and traveller Brian Blessed spoke of the capture of a similar creature in Russia, lending an air of realism to the Fouke Monster discovery and legend. He said:

"...*Russian evidence is pretty sensational, in a town called Takeema at the end of the last century, they caught an Almas giant, a woman, about seven-and-a half feet tall called Zana. She had lots of children. Wicker's World met some of the grand-children who were enormous. Zana was almost eight-feet tall, covered in hair...they kept her in cages, they kept her in a stockade. She became part of the village, she ploughed the fields. And she had children, and bred with human beings."*

If a similar animal inhabited the Fouke swamplands and existed as a real specimen, why did it seem immune to gunfire? There were reports that a similar creature was shot at in 1963, and what with the James Lynn encounter, plus numerous other, if debatable accounts, it would seem that the monster should have perished a long time ago. That is if the local people were sure-shots. Whilst many so called bigfoot investigators believe such creatures to be almost supernatural, I do not want to bring the Fouke Monster into any such realm. This thing was a flesh and blood animal that perished, leaving behind the greatest evidence for the existence of such a creature.

According to other sources, knowledgeable or not, many other sightings of the Fouke creature have taken place. In 1978 alleged footprints measuring 17 inches long and 7 inches wide were discovered in Appleton near so-called caves, in Arkansas. Ten miles away at Centre Ridge more prints were found and livestock reported missing.

In Crosset during the summer of the same year, Mike Lofton supposedly shot at a seven-feet tall 'something' which lurked outside of his house. Mike was feeding his puppy dog when he noticed the creature emerge from woodland. It is only when we learn that Mike was ten-years old that the encounter becomes debatable. According to reports the child retrieved his father's .22 calibre rifle from the house, approached the animal and blasted it.

During the early 1980s, a family living fifty miles east of Hope, near the Ozark Mountains and Black Lake, on several occasions sighted a seven-foot tall creature which left 17-inch long footprints. The animal was described as, "…gorilla-like…more human than animal…its arms much longer than a man and the face covered in hair".

Allegedly dermal ridges were seen on the prints left by the creature and there was never any indication of an arch. Such an account contradicts the evidence found in the Fouke soybean field. Don Pelfrey, of Kentucky, whose relatives saw the creature at Hope, believed the animal to weigh in the region of eight-hundred pounds and to stand ten-feet tall. Once again, nothing like the original Fouke Monster, which was estimated to weigh between two and three- hundred pounds!

During 1977, a number of animals were found mutilated on a farm belonging to the aunt of Don Pelfrey. Dogs, one calf and a number of chickens were found in a mutilated state. Two large hogs were reported missing one night and discovered the next day some five-hundred yards away. The innards had been torn out and scratch marks were apparent. According to sources, the hogs had been killed for sport rather than food. Nearby, a neighbour had two Doberman dogs killed. They were found mutilated and crushed.

According to newspaper *USA Today*, Fouke remains in the top ten list of places to visit with regards to bigfoot. Back in the '70s the mystery lost all reason as the local sheriff's office stopped visitors in their cars in order to check for guns and booze and ordered them not to trespass. Of course, much of this went through one ear and out the other as poor old Smokey Crabtree had to construct signs to keep people off his premises. It never worked.

Little Rock radio station, KAAY, put up a reward of $1,090 to anyone who could dig out the creature. This was not a good idea. Various arrests were made on hoaxers and drunks rampaging through backyards, woodlands and fields, creating chaos.

Like all great enigmas there is always the essence of some great conspiracy or local law enforcement cover-up. There was previous mention that some within the Fouke community knew where the creature had come from but former sheriff Les Greer said the sightings went way back to 1946. If this is the case then the travelling circus story seems a cover-up for either more sinister events or a mere misinterpreted item of folklore.

Greer was sheriff of Miller County from 1967 to 1974. He heard about the creature from a local lady whilst campaigning for tax assessor. He said:

"She lived about halfway between Fouke and the Below Bridge. She told me that she saw some kind of animal go down in the field in a low, bushy place. She said it looked kind of like a man, and walked like a man but she didn't think it was a man!"

Fouke Mayor Cecil Smith was sceptical of the monster's existence.

"I used to racoon hunt down there and people would ask me if I was scared but I never felt like anything that big could manoeuvre around."

Opinion seems divided with regards to not only the creature's existence but to what it actually was. Indeed, as is always the case, no-one has ever seemed willing to grasp the legend fully, instead the sensationalism comes to the fore, creating hysteria and confusion. Newspapers and locals find no interest in the possibility of *Gigantopithicus Blacki,* a giant orang-utan which existed some five-hundred thousand years ago, roaming Fouke, or other relic hominids, instead they thrive on tales of unknown monsters and money-making schemes. Such trivial falsity may well enable the legend to live forever but the scientific and cryptozoological truth remains clouded.

THE SKELETAL ANALYSIS

Smokey Crabtree: *"The first time I heard about it, it was found over on the edge of Texas... some boys had found the skeleton of the Fouke Monster."*

The 1996 video *"The Hunt For Bigfoot"* seems like any other monster or unexplained mystery documentary...except for the fact that it goes beyond the atmosphere and speculation to display raw reality. Whilst many have scoffed at the famous 'alien autopsy' movie or still ponder over the Patterson-Gimlin Bluff Creek Bigfoot, *"The Hunt..."* zooms in on something potentially startling, that has, for the moment anyway, remained hidden.

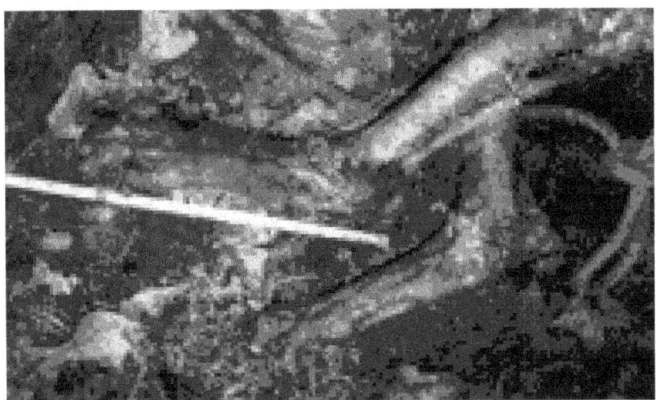

Skeletal legs of Fouke monster found in 1991

According to Smokey the remains of the 'creature' were still blood-covered, the corpse didn't stink and there were patches of hair. Whilst much analysis has been applied to the skeleton the main mystery lies in the undisclosed information regarding DNA tests, hair analysis and the lack of information since 1996.

The documentary gives mention to at least twenty other sightings of the Fouke creature whilst opinion within the community seems divided although fascinated. The main frustration with the video of the remains is down to the forty-five minute running time and the absence of certain measurements and analysis whilst other details are included. However, expert comment seems promising.

Dr. Beth Leuke (Professor of Biology): *"The foot bones are very strange looking. Ribs are definitely collapsed. That is not a human pelvis bone, it is too long. The hands and very feet are very primate looking... long fingers, gorilla-like. The head was removed from the body easily but I would have been surprised if wolves had taken it. Judging by the size of the body the skull would have been thick and heavy and difficult for wolves to get into. Most scavengers go for the viscera... the gut, then dismember arm bones and leg bones. In this video all appendages are present so I have a hard time believing scavengers have been to work on it."*

Dr. Vaughn Langman (Professor of Biology): *"...certainly big feet, it doesn't walk on its digits, it walks flat-footed. A very strange skeleton..."*

Richard Manning (Taxidermist): *"I don't know of any animal in thirty-two years that I have skinned that had a heel like that on it. The legs have to be four-feet long. It is humanistic but more like a gorilla. It is not a cat or bear because of the heel."*

The remains, in a glass case, were loaded and taken to a Louisiana for further analysis from a Forensic Anthropologist. Clay Stewart:

"There are inconsistencies which tell me it is non-human remains. First of all, the scapular, or shoulder blade of a human is much smaller. The broad blade of this scapular and large muscle attachment is inconsistent with that of a human. If we look at the vertebrae, an entire human backbone measures about nineteen or twenty inches in length, as opposed to the vertebrae here which measures fifty inches. Another inconsistent detail is the pelvis. The pelvis is very long and narrow and the human pelvis is short and broad, very rounded and shallow."

Further analysis by a team of University Biologists revealed the following measurements:

- Legs – Femur (35 cm – 14 inches)
- Shin Bone (36 cm – 14.5 inches)
- Feet - 32 cm (12 inches approx)
- Strangely, the legs of the 'creature' are human-like in length.

Judging by the photograph it would seem that the creature *does* walk on its toes especially as its foot is nothing like the typical sasquatch flat print. After the remains are seen the Fouke monster legend becomes inconsistent, unless of course we are dealing with several species of unidentified primate. Whilst the remains and their analysis have to exude more information, one cannot help but feel that the creature that once walked in that structure was some sort of giant monkey, a creature identified in Loren Coleman and Patrick Huyghe's *"The Field Guide to Bigfoot, Yeti, and Other Mystery Primates Worldwide"*. The authors state clearly that such creatures are widespread, but more likely to inhabit Asia although North America has its own three-toed 'Devil Monkey', a six-foot tall muscular creature said to resemble kangaroos! However, a quick glance at the skeletal remains and you would be forgiven for thinking of such an animal with its protruding heel. However, the Patterson-Gimlin was not kangaroo-like at all and also had a large heel to support its bulk.

Footprints of the giant monkey are said to reach twelve inches in length and the behaviour is known to be hostile, especially towards man. Blood-curdling screams are also said to soar from the monkey whilst the dreaded call of the Fouke beast was said to start as a deep bellow and then become more of an upbeat racket and then a scream, as if the creature was demanding and claiming territory.

Although fossils of a giant howler-spider monkey have been discovered in South America it would seem more likely that the local community has been home to one of Coleman and Huyghe's 'unknown pongids', a species of unknown primate which covers large orang-utans and skunk apes. Such creatures are said to reach eight feet in height, have short legs and long arms, a small head and are covered in shaggy hair. However, the confusion arises with regards to the feet. The pongid's pads not only long but wide and displaying five toes, so if the Fouke Monster prints were real then its characteristics differ.

According to the Coleman/Huyghe book, *"Unknown Pongid's"*, Marked Hominids and Erectus Hominids were all sighted in the swamplands of Texas, Minnesota, Missouri and Arkansas during the late 1960s and early '70s. Were all these the one species of Fouke Monster or a variety as stated? A shaggy beast seen July 11th 1972 in Louisiana by three witnesses stood seven-feet tall, omitted a terrible scream and gave off a terrible odour.

Two years previouslt, in February, a male witness saw a five-and-a-half foot tall creature in a tree at Central, Arkansas. The animal leapt from the tree and swung its arms like a gorilla before making a deep breathing row. This encounter though was unlike other Arkansas sightings though, this time the creature seemed to be covered in tight fur like a hound and have a balding head. In the same area around the same time there were numerous reports of chimpanzee-like animals and encounters with gorillas. Were all these accurate descriptions or misidentifications which is a detail currently attributed to mystery feline sightings and lake monsters where, despite the credibility of the witness, it seems as though they are not quite sure what they have seen. This is not an air of scepticism, just a fact.

And so, the legend and possible truth of the Fouke Monster lies inside a glass case somewhere in the middle of nowhere. Like so many other great mysteries it will leave us all bereft of a much-needed update. Like the lasting Patterson-Gimlin film the aura is there but the scientific facts are not.

Many witnesses just call it bigfoot, whilst the scientific community fail to put a name to it. Both parties are clearly wrong. Whatever is, or was out there was never that elusive due to the number of sightings, but it was intelligent enough to create a legend as everything snowballed around it. The pile of bones in that case may well be the greatest zoological, anthropological and scientific discovery ever, or they may just be another unsolved mystery that often pops up when one is busy dissecting another.

And so we go back to "The Hunt For Bigfoot" and the words of a Dr. David Otto:

"If we found a bigfoot we would probably find a way to use our science, culture and arts to dismiss it quickly. We would make it less significant than we once thought it would be and suddenly something much larger than bigfoot would come along...a bigger foot still. We need bigfoot to help us as k some fundamental questions about who we are as people, who we are as individuals and what it means to deal with that wild side within all of us."

Or in the immortal words of Smokey Crabtree.

"I've got no idea...!"

Morgawr the Sea Dragon

An Overview of twelve years of research

by Jonathan Downes

NOTE: In the six years since I wrote *The Owlman and Others* new information has come to light. This extract from my forthcoming book *Monster Hunter* brings the story up to date

Unlike many of the places mentioned in this book, some of the areas where I have travelled and which I have explored looking for monsters are easily accessible to the resident of the United Kingdom. Forgive me, therefore, if I follow my descriptions of high strangeness in the Lesser Antilles with a few paragraphs of Tourist Information on how to find the places described in this next chapter!

Drive to Exeter, either by using the A303, or by the M5, and then get on the A30 heading west. As soon as you enter Cornwall, it is like you are entering a foreign country - indeed according to many Cornishmen and women, Cornwall - the ancient land of Kernow - IS a different country. Carry on driving west down the A30 until you see the signs for Truro. Drive through Truro following the road west for Falmouth. When you actually get there drive up the hill and follow the signs for the sea front and docks rather than for the town centre. Drive across several roundabouts and then down a hill following the signs to the docks until you reach a large and dull municipal carpark to your right, and the legendary Pirate Inn (home of Cornwall's best live music venue) on your left.

Drive slowly up the crowded street, and turn sharp right down a steep lane leading to 'Customs House Quay'. Find somewhere to park and gingerly approach the pub called 'The Chain Locker' which you will see on your left. . .

If, as you enter the bar, you see an imposing Irishman with a black beret and a bristling grey beard, buy him a pint of Guinness. If he is talking to an overweight bloke with glasses and a slightly less impressive beard buy me a drink and tell me how much you like this travelogue. If you don't like it keep schtum or you may get the combined wrath of a scorned author and the wizard of the western world.

I have given fairly detailed instructions about how to get to this particular pub because the pubs of this picturesque and bustling seaport were where I carried out most of my investigations into Morgawr – the not so legendary Cornish Sea Serpent, under the tutelage of my sometime mentor Tony "Doc" Shiels.

Tony is unlike anyone else I have ever met. My ex-wife once said that meeting him for the first time was like being struck by a whirlwind. Tony was born in 1938; and, over the next fifty-seven years, he would be painter, conjurer, gunslinger, musician, playwright and husker. He is also a self-admitted wizard. He has never claimed magical powers of his own; but, then again, he has never claimed not to have them either. My own feelings are that Tony is a very similar character to Reg the time traveller in Douglas Adams's *Dirk Gently's Holistic Detective Agency*. Reg discovered the secrets of time travel by accident because he could never be bothered to learn how to programme his video recorder, and who could never figure out alternative ways he could see episodes of TV shows he would otherwise have missed.

I think that Tony developed his very real magical skills because he could never figure out how to hide the hard-boiled egg up his sleeve like a proper conjuror He is shorter than I (as are most people - but, then, I am six-and-a-half feet tall), and has piercing, powerful eyes which twinkle when he is amused, and cut like a laser beam when he is annoyed; short-cropped, grey hair; and an enormous, bushy beard which bristles magnificently in all directions. He comes over like a cross between a genial Mephistopheles and Captain Birdseye with a cosh in his pocket. He drinks Guinness and smokes small cigars. For nearly a decade now I have been proud to count him as a dear, if somewhat unpredictable, friend.

The good doctor, whom an Irish Newspaper dubbed "The Wizard of the Western World" at the height of his infamy during the late 1970s was, as we shall see, pivotally involved in the Morgawr story, although as we will also see, the accounts of a giant sea monster off the Cornish coast date back for hundreds of years.

Tony Shiels is not the only wizard with whom I became involved during my researches into the Cornish sea monster. I share a house with one of my colleagues from the Centre for Fortean Zoology. His name is Richard Freeman and he joined forces with Graham and me after he finished his degree in Zoology at Leeds University. Richard Freeman is not just my housemate and fellow toiler in the Cryptozoological vineyard, but he is a Goth with the predilection for high strangeness that often accompanies a taste for doomy music and stupid haircuts. He is also an accomplished occultist and a practising wizard, who although not aspiring to the lofty heights of someone like Tony Shiels, is not unskilled at various branches of the arcane arts.

During my chequered career as a monster hunter I have found myself in some fairly strange places, but it is unusual even for me to find myself on a southern Cornish tourist beach in the middle of winter accompanied by a TV film crew, a minor local celebrity turned TV presenter, my dog Toby and two of my colleagues from the Centre for Fortean Zoology.

It was only about ten days after Graham's and my return from Mexico and it was a typically cold and grey morning in early March. Graham and I were shivering as we daydreamed wistfully about cold beers under the desert sun and did our best to keep out of the sleety wind, which cut through us like a knife. Toby, by then quite a venerable old dog at the age of thirteen was still fairly spry and he wandered happily about the beach, trying to eat seaweed and cocking his leg against the jagged rocks oblivious to all that was happening around him.

Apart from Toby, everybody else was staring awe struck at Richard who stood, legs akimbo, impressive in a long black robe and brandishing a fierce looking sword towards the sea. He was chanting an ancient invocation in a mixture of Gaelic and Olde English in an attempt to summon the ancient sea beast from its lair.

Around him on the shell strewn sand were four candles marking the cardinal points of the sacred circle that he had cast. They spluttered bravely against all odds, and I am sure that everyone present was amazed that they had not been extinguished by the sleet and the wind. Each of the candles were different colours: green for the north, signifying the element of Earth, red for Fire marked the southernmost point of the circle, a yellow candle signifying the element of Air stood in the east, and the western point was marked by a blue candle representing the element of Water.

As his invocation reached its peak, Richard screamed the ancient Gaelic incantation at the unyielding sea and threw a bunch of elderberries wrapped in grey cloth into the waiting sea as a gift to the Cailleach [1]

Suddenly he started to scream, in English *"Come Ye Out Morgawr, Come ye out ancient Sea Dragon, Come ye out Great Old One"* and as he screamed his face began to change shape, and the friendly happy bloke that I have known for years and who currently resides in my spare room imperceptibly seemed to change into something darker, stranger and as old as the Cornish landscape itself.

Without even realising it all the people present turned as if synchronised towards the sea, all of us expecting to see the dark head and long neck of an ancient sea dragon rise from beneath the slate grey waters…

Richard was putting his own spin on something that Tony Shiels had done back in 1976. In late March 1976 Tony and a colleague called a press conference in *The Chain Locker*. According to the local newspaper report of the time:

"The great Falmouth monster hunt got underway this week, with the arrival in the town from America of a 36 year old Professor of Metaphysics, Michael McCormick, who plans to devote the summer to his search. The professor called a press conference in the Chain Locker on Falmouth's waterfront on Monday, when he was accompanied by his `psychic advisor`, `Doc`

Shiels of Ponsanooth. He had with him a number of strange relics from previous monster hunts, including the skeleton of an imp. This was about eighteen inches long, with a miniature human shaped skull out of which horns protruded. It also had clawed feet and wings, and did not smell too wholesome.

He also produced a small, clawed foot, wrapped in a red cloth, which promptly caused oscillations on the sound recording equipment brought along by television engineers.

Prof. McCormick said he had recently been lecturing at the University of New Mexico on the basic need for monsters from a Jungian point of view. He said he had come all the way from Alberquerque, New Mexico, as a result of reports in the `Packet`, about sightings of the Falmouth monster. `It has cost a wad of money`, he said, `I shall spend the next two months in a determined focussing effort using "Doc`s" remarkable mental processes to produce the beast`. He said that he thought the monster was probably migratory, and thought it could vary its size at will. The professor, who is also a fire-eater and juggler, is causing some problems for his partner. Said Mr. Shiels, `The whole place is throbbing from the things Mike has brought over. I shall have to do something to stop the headaches`. " [2]

According to students of the ancient, and now practically extinct Cornish language the word `Morgawr` means 'Sea Giant', and was used to describe an enormous marine monster which has always been said to live in the waters of Falmouth Bay. According to an 1876 newspaper report:

"Port Scatho. The sea-serpent was caught alive in Gerran`s Bay. Two of our fishermen were afloat overhauling their crab pots about 400-500 yards from the shore when they discovered a serpent coiled around their floating cork. Upon their approach it lifted its head and showed signs of defiance upon which they struck it forcibly with an oar which so far disabled it as to allow them to proceed with their work after which they observed the serpent floating around near their boat. They pursued it, bringing it ashore yet alive for exhibition soon after which it was killed on the rocks and most inconsiderately cast into the sea" [3]

Author and Zoologist Michael Bright is amusingly sceptical about this incident. In 1989 he commented *"How anybody can continue with their work after an encounter with a strange sea creature beats me. And the report does not say what happened to the creature after it had been 'cast again into the sea'.* " [4] and I have to say that I cannot help but agree with him. From the description, the creature was probably an enormous eel of some description.

There have been accounts of giant marine eels along the channel coast of the South Western peninsula for many years. Richard Freeman is not just a magician, but also a zoologist and he has collected many such accounts for the CFZ archives.

As a boy in the '70s Richard holidayed in Devon. One summer, when he was aged about 9 his grandfather got talking to a retired trawler man in Goodrington harbour. The old man recounted his life as a fisherman and one particular incident that had stuck in his mind. Some years previously he and his crew were trawling off Berry Head. The seas off this part of the coast are amongst the deepest waters around Britain. Such are the depths of this part of the

English Channel that the area is commonly used to "scuttle" old ships. The drowned wrecks of these vessels have made an artificial "reef" that has attracted vast amounts of fish. Good catches were therefore almost guaranteed and the area has become a popular place to drop the nets.

One night the crew had trouble lifting the nets and began to worry that they had got them entwined about a rotting mast. Soon though they felt some slack and began to haul the nets up. The men thought their catch must have been a particularly good one so heavy were their nets, but as their nets drew closer to the trawlers lights a frightening sight took shape. The crew had not caught hundreds of normal sized fish but one gigantic one.

"It was an eel, a giant eel. Its mouth was huge wide enough to have swallowed a man, the teeth were as long as my hand." Even now Richard still remembers the words of the ancient fisherman and is convinced that this was not a tall story designed to entertain gullible tourists. *"While it was still in the water it was buoyed up but as soon as we tried to pull it on board the nets snapped like cotton and it vanished back down. I was glad it went, I've been at sea all my life but I've never been as scared as I was that night. I can still see it's eyes, huge, glassy. . . . . "*

The man couldn't see what length the beast was as it had been coiled in the nets. We can however make a guess. We don't know what the breaking strain is on the average trawlers' net but logic dictates that it must be several tons, so this must have been a truly massive animal. In order to have a mouth wide enough to swallow a human, a standard conger eel would need to be scaled up to around 50 feet!

This was only one of the legions of such tales from Devon's rocky coastline. In the late 30s a creature described as a "giant conger eel" terrorised the south coast of Devon, frightening fishermen and tourists from Berry Head to Plymouth. Events such as this clearly show we cannot confine monsters and dragons to storybooks and campfire tales.

However, not all the historical accounts of Morgawr appear to refer to giant eels. Fifty years after the unfortunate anguilliform was so summarily slaughtered in Gerran`s Bay Mr. Reece and Mr. Gilbert, trawling three miles south of Falmouth netted an amazing creature. It was twenty feet long, with an eight-foot tail, a 'beaked' head, scaly legs, and a broad back covered with 'matted brown hair'. Marine Biologists of the day were unable to identify the beast.

Again, unfortunately for those who would like to claim that all of the historical accounts of unidentified sea creatures off the Cornish coast refer to species presently unknown to science, it is fairly easy to identify this beastie as well. As long ago as 1968 Professor Bernard Heuvelmans noted that:

"It seems to be quite normal for basking sharks to take on a plesiosaur's shape when they decompose. This is because of the peculiar structure of their gill slits, which ate extremely long and go almost right round the neck. A live basking shark is virtually decapitated. As soon as its tissues decompose and become soft, the whole gill apparatus easily falls away taking the

jaws with it, so that nothing remains in front of the pectoral fins but the tiny skull and the spinal column clad in its muscles and thus looking like a thin neck. At the other end of the body, the lower fluke of the tail soon goes with nothing to support it, since the spine extends only into the upper one. The body then seems to have a long thin tail.

The final monstrosity of these pseudo-plesiosaurs is that they seem to be hairy. in the selachians the fibres of the surface muscles break up into whiskers when the skin rots or is eaten away. These fish then seem to have coarse stiff fur, varying from dirty white to reddish in colour, as the body decomposes or dries out on the shore. if parts of the dorsal fin remain these hairs can even look like a mane. Thus there is no need to be as bold or simple as a mediaeval artist to produce a creature with a fish's skeleton, the general shape of a reptile and with mammal's hair. " [5]

Unfortunately the newspaper reports of the time are unclear whether the creature found by Reece and Gilbert was alive or dead. However, if, as seems likely, the creature was dead when they hauled it in, one need look no further for an explanation of its provenance than Heuvelmans's masterful description. Furthermore, providing even more bad luck for those who are truly in search of an antediluvian monster Heuvelmans went on to describe another strange carcass which was washed up on Prah Sands in southern Cornwall in 1933 and states that he believes that this animal too was a decomposed Basking Shark.

However some Cornish monster sightings are far less easy to explain. In 1949 author Harold Wilkins was in the Cornish fishing village of Looe when they saw:

"Two remarkable Saurians, 19 - 20 feet long, with bottle-green heads, one behind the other, their middle parts under the water of the tidal creek of East Looe, Cornwall, apparently were chasing a shoal of fish up the creek. What was amazing were their dorsal parts: rigid, serrated and like the old Chinese pictures of dragons. Gulls swooped down towards the one in the rear. These monsters -and two of us saw them - resembled the plesiosaurus of Mesozoic times. " [6]

It seems that there really were some strange creatures in the briny depths of Falmouth Bay because one sunny evening in September 1975, Morgawr was spotted off Pendennis Point. Mrs Scott, of Falmouth, and her friend Mr. Riley, saw a hideous, hump-backed creature, with 'stumpy-horns', and bristles down the back of its long neck. The huge animal dived for a few seconds, and then resurfaced with a conger eel in its jaws. Mrs. Scott told local journalists at the time that she would never forget 'the face on that thing', as long as she lives. [7] Shortly after the Scott/Riley sighting, Morgawr was encountered by several mackerel fishermen, and blamed by the superstitious fishermen for bad luck, bad weather and bad catches.

In February 1976, the *'Falmouth Packet'* newspaper published two photographs of Morgawr, taken in February by a lady who called herself 'Mary F'. They showed a long necked, hump backed creature, at least eighteen feet long, swimming in the water off Trefusis Point, near Flushing. A letter accompanied the photographs:

"The enclosed photos were taken by me about three weeks ago from Trefusis. They show one of the large sea creatures mentioned in your paper recently. I'm glad to know that other people have seen the giant brownish sea serpent. The pictures are not very clear because of the sun shining right into the camera and the haze on the water. Also I took them very quickly indeed. The animal was only up for a few seconds. I would say it was about fifteen to eighteen feet long. I mean the part showing above the water. It looked like an elephant waving its trunk, but the trunk was a long neck with a small head on the end, like a snake's head. It had humps on the back, which moved in a funny way.

The colour was black or very dark brown and the skin seemed to be like a sea-lion's. My brother developed the film. I didn't want to take it to the chemist. Perhaps you can make them clearer. As a matter of fact the animal frightened me. I would not like to see it any closer. I do not like the way it moved when swimming.

You can put these pictures in the paper if you like. I don't want payment, and I don't want any name in the paper about this. I just think you should tell people about this animal. What is it?

Yours sincerely
Mary F"

These photographs are, with the possible exception of the 1967 Roger Patterson/Bob Gimlin Bigfoot film, the most contentious images in contemporary forteana. They polarised the fortean establishment with many people staking their reputations on the veracity or otherwise of these two rather dubious images. One of the people who came out most strongly on the side of those who believed it was a hoax was veteran American fortean Mark Chorvinsky. He devoted an entire issue [8] of his excellent *Strange Magazine* to debunking them, and to exposing Tony "Doc" Shiels as the perpetrator of the hoax.

In order to substantiate the claim he relied strongly upon evidence given by several of Tony's friends *including* the aforementioned Mike McCormick and some appallingly crude photographs published by Tony in a 1976 book in which he explained various techniques for producing a 'Psychic Superstar' by media manipulation. These techniques included the hoaxing of monster photographs.

The photographs were credited to 'G. B. Gordon', and Mark Chorvinsky announced with a flourish that they were actually fakes by none other than Tony himself! Of course they were. Tony has never denied this. They were never meant as anything apart from slightly amusing illustrations for a book which was itself an amusing spoof on a bestseller by Uri Geller. [9]

In my 1997 book *The Owlman and Others* I described how in my opinion Mark Chorvinsky had developed somewhat of an *idee fixe* about Tony Shiels and his propensity for playing the psychic prankster. I maintained then, as I do now, that Mark is a fine researcher and someone for whom I have nothing but praise. Unfortunately I believe that the whole affair of *Strange Magazine #8* was and is Tony Shiels's greatest ever hoax!

In recent years, and certainly since the early 1990s Tony has been keen to be seen as an artist rather than as a fortean and he has done his best to play down his monster-raising past. He has featured in a number of high profile art exhibitions and his surrealist paintings, drawings and sculptures are at last beginning to achieve the fame, and the prices that he has deserved for so many years.

He has been a painter for nearly half a century now, and he has told me on a number of occasions how he is tired of only being known to the general public as a genial eccentric who *"messes about with sea serpents"*.

I believe that when Mark Chorvinsky first approached Tony and his compadres for information about the genesis of the Mary F photographs and other monster pictures allegedly associated with Tony over the years, the shamrock shaman saw his chance to redress the balance.

Chorvinsky interviewed a number of people on the subject, and various people from his past, like McCormick, Roy Standish - an ex-editor of the *Falmouth Packet* - and Alistair Boyd, the researcher who finally uncovered the truth about the infamous "Surgeons Photo, " which for well over half a century was seen as the most convincing piece of evidence of the Loch Ness Monster.

They painted a convincing picture. Roy Standish said that even at the time, he had been convinced that Doc was responsible for the Mary F photographs, and several people alleged that Doc had sent them prints which were slight variants on the Mary F photographs several weeks before the images we have come to know appeared for the first time in the *Falmouth Packet*.

This was probably true. Tony even admits as much. He claims, (and knowing the way the Fortean underground works I tend to believe him), that several different versions of the photographs had been circulating around the Falmouth area before they were actually published.

This makes sense, and although by implication the story given to the *Falmouth Packet* with the photographs is probably untrue, there is no real reason to suppose that Tony is responsible for the photographs.

The main problem with viewing the Mary F photographs as a piece of art by the great surrealist painter Tony Shiels is that they are actually not terribly good, which is something that no one could say about his other pieces of work. They are spectacularly unconvincing pictures which I have no difficulty in believing were made by using a plate of glass and some modelling clay. I just have a sneaking suspicion that if Tony had produced them, they would have been done much better.

I said as much in *The Owlman and Others,* but I stressed that although I was, and still am convinced, that poor old Mark Chorvinsky was well and truly bamboozled by Tony Shiels, that there is no shame in being bamboozled by such a master of the art as the good doctor. He has pulled the wool over my eyes on various occasions and I believe that the only reason that I managed to come to such diametrically different conclusions in my research than Mark did in his was that I had access to far more of the original source material, and that I was lucky enough to spend an inordinate amount of time drinking Guinness around Cornwall's pubs with the man himself.

Here, I should perhaps include a note on Cornish pubs for the aspiring fortean investigator. If you visit one of the more traditional hostelries and order a pint, you can drink about a third of it, and then ask for a `Cornish Half'. They will then top it up for the price of only half a pint. Not all pubs in the county do this, but by asking for one you immediately establish yourself as someone 'in the know' which may prompt the more taciturn of the locals to be more voluble on the subject of sea serpents and owlmen!

There is no doubt that the Mary F photographs are fakes. The only serious doubt is over the matter of who took them. Forteana is a very strange business by its very nature, and Tony Shiels told me on a number of occasions that there really isn't any such thing as a coincidence. Therefore, I was saddened, though not particularly surprised when, as I was putting the finishing touches to this chapter in late January 2001, I received a letter from Tony telling me that the person that I had always suspected has been the real hoaxer of the Mary F photographs had recently died. His name was John Gordon, and during my long and tortuous researches into the truth about what happened during the long, hot summer of 1976 at least three people hinted strongly that John had been the real perpetrator of the Mary F hoax.

I hinted as much in *The Owlman and Others* but as my informants had been speaking strictly off the record, and also because, at that time John was still alive I was not really in a position to reveal the fact. Now he is dead it seems a reasonable premise to do so.

The important thing is that Tony did not fake the pictures himself. They are fairly unimportant as fortean artefacts unlike his later pictures of Irish lake monsters and even Nessie herself. I believe that to cast doubt on Tony's veracity *vis a vis* the Mary F photographs is dangerous to forteana as a whole because it then brings the results of Shiels's very real contribution to contemporary forteana into jeopardy.

I was not prepared, however, for the storm of controversy that would occur when Mark Chorvinsky was to read my book in which, after all, I had hailed him as a fine researcher for whom I had nothing but respect. I met Mark at the 1998 *Fortean Times* Unconvention where he and I were both speaking. I autographed a copy of my book for him and told him about my adventures with the more recent monster raising invocatory activities a few weeks earlier.

We parted on good terms, and I went off to my hotel room with my then girlfriend for some illicit lechery and to plunder the mini bar. The next morning Mark Chorvinsky was livid. He hated the book and took grave exception to everything I had written about him. He seemed to think that somehow I was casting doubts on his ability and professional reputation as an investigator. *"Of course I'm not, dear boy"* I said in the calming Old School voice which usually manages to soothe our transatlantic cousins when their feathers have been ruffled. *"I'm merely reaching a different conclusion to you"* … but it was no good.

On his return to the United States he sent vitriolic e-mails to various correspondents, including, I believe some Internet discussion groups, and from what I can gather told everyone that he could that he regarded me (and presumably still regards me) as some sort of fortean antichrist for daring to claim that he might have been mistaken.

I couldn't be bothered to join in the mudslinging because not only was he not the first person to attack me in print that year, and he certainly wouldn't be the last, but I really didn't care. After all, despite the highly dubious provenance of the Mary F. photographs there was still a reasonably good body of evidence to support the hypothesis that there was indeed some strange creature at large in the waters of Falmouth Bay. Even more astounding was the evidence suggesting that somehow, it was indeed linked with some kind of magickal energy and that Tony "Doc" Shiels was capable of summoning the beast under certain circumstances.

In December 1976 Tony Shiels was on the shoreline below Mawnan Old Church with David Clarke from *Cornish Life* magazine, who was doing a photo shoot on him. He wanted photographs of Tony invoking Morgawr to illustrate a feature he was planning. Tony duly obliged, and David Clarke was taking photographs when:

"I saw a small dot moving towards us, which I presumed to be a seal. It came across the river to within 60-70 feet. It started to zig-zag backwards and forwards, and I could see movement in the water well behind the head, which suggested that it was a great deal longer than a seal." [10]

Clarke's dog started to bark at the animal, and it sank from view. He was immediately suspicious of Tony, and accused him of setting up the illusion as some kind of trick. This is something that Tony has always strenuously denied.

The interesting thing is that while Tony's pictures, which were taken on an inferior camera without a telephoto lens, came out properly, although the image was indistinct and not really conclusive, the "jinx" which has bedevilled Fortean researchers for many years, and in many continents, struck again; and David Clarke's camera malfunctioned, causing pictures which were seriously double-exposed. They were, however, probably the most convincing pictures yet obtained of the mysterious creature of Falmouth Bay.

It should also be remembered that whilst such respected authors as Michael Bright, and practically every other author who had written about the subject of British sea serpents had quoted Mawnan-Peller who claimed that:

"In January 1976 a strange (and, so far unidentified) carcass was discovered on Durgan Beach, Helford River, by Mrs. Payne of Falmouth. For a while it was thought that the monster was dead" [11]

Everyone in the field were overjoyed at the discovery of this wonderful piece of evidence and nobody actually thought to investigate it further, or to ask Tony Shiels himself about it.

By the time that I was first investigating the Morgawr phenomenon in Cornwall during the summer of 1994 the belief that there had been a genuine monster corpse washed up rotting on a Cornish beach, and that only stupidity on behalf of some un-named `powers that be` had prevented science from getting their sticky fingers on a wonderful piece of cryptozoological evidence had become entrenched in the canon of fortean belief. Like so many entrenched beliefs, fortean or otherwise, the truth, though far stranger in some respects, was far removed from what had been written in so many books and magazines on the subject.

A few weeks after the initial discovery of the headless carcass on Durgan Beach, the *Falmouth Packet* reported that a young naturalist living locally believed he had found the answer to the mystery:

"The mystery of the bones of Durgan Beach may have been solved this week by 13-year-old Toby Benham, a keen student of skeletons. Toby believes the bones found at Durgan by Mrs Kaye Payne of Falmouth, come not from a 20 foot sea monster as suggested, but from a whale.

He came to this conclusion because he thinks that the bones form part of a skeleton he discovered on nearby Prisk Beach just after Christmas. Toby studied the Packet's photograph of Mrs Payne holding a bone from the beach, and he is convinced that it is one of those he saw.

"I am sure it is from a whale," said the young naturalist emphatically. His explanation for their appearance at Durgan is equally emphatic. Storm tides swept them around from Prisk, he says. The original skeleton was about ten feet long, and the skull, which is now one of the prizes in Toby '5 collection of bones, looks like that of a whale. He said the skull had what appeared to be blowholes and it seemed very similar to pictures he has of whales' heads. " [12]

Unfortunately, although several books report Mawnan-Peller's description of the finding of a "carcass" on the beach; none, that I have been able to find, have reported the fact that the "carcass" was actually a headless skeleton, and that there is every likelihood that the skeleton itself was actually that of a whale. Only one book actually acknowledged the whale theory, and that was *Monstrum* by Tony "Doc" Shiels; and, for reasons known only to themselves, the fundamentalists among the world of Forteana have decided to take it upon themselves to ignore everything that Doc ever says!

"Of course it was a bloody whale" Tony blustered at me one night as we were sitting around the kitchen table drinking rum out of coffee mugs. *"Here, I'll show you"* and he rushed away from the table into the dark interstices of the next room from whence he emerged a few minutes later clutching a colour transparency showing a pretty teenage girl clutching what was undoubtedly one of the vertebrae of a dead whale. *"My daughter Lucy"* he barked at me, before changing the subject and demolishing the rest of the bottle of rum.

However, having been caught out once already by the oft repeated assertion that the carcass was indeed that of an unknown animal we embarked on a quest to see if we could get hold of Toby Benham and the elusive whale skull. By dint of some major detective work we managed to track down Toby's mother who said that, unfortunately, Toby was no longer living at home. As, by this time he would have been nearly as old as me this did not come as any great surprise, but we asked diffidently, whether there was any chance that his mother knew what had happened to the elusive skull. Of course she did, she told us. When Toby had grown up and fled the nest he left the skull behind, and for many years it had been lying open to the weather in his mother's garden where it doubled as a door stop and a somewhat macabre garden ornament. After some years she had got tired of it and donated it to a local educational institution.

Three weeks later in a dusty cupboard of a locked classroom in a local college this is where we found it. The proprietors of the college were happy to let the CFZ have it as a specimen for our nascent collection of cryptozoological memorabilia, and so we carried it gingerly out to our van and took it home which is where it resides to this day. We had it identified by an expert at Plymouth Aquarium and by Dr Karl Shuker who both correctly stated that it was the somewhat damaged skull of a baby pilot whale thus vindicating what Tony Shiels had stated all those years before.

Many people have claimed that all the sightings of the Falmouth Bay monster can be directly linked with Tony Shiels. This is clearly not the case.

In 1996 local author Sheila Bird wrote a letter to the *Falmouth Packet*:

"For anyone who has actually experienced a sighting of the strange marine creature which has popularly become known as Morgawr, the monster of Falmouth Bay, japes and jokey reports of sightings are particularly frustrating in that they discredit genuine reports and reporters, and thereby discourage serious investigation which could possibly lead to a scientific breakthrough.

This being the case I have decided to place on record my sighting of this creature off Portscatho on 10 July 1985, which I did not report at the time for it coincided with the launching of my book Bygone Falmouth, and I was reluctant because it might have been interpreted by some, as a joke to create timely publicity.

Having just met my brother, Dr Eric Bird of Melbourne University, who is a scientist and who had flown in from Australia on that day, we were relaxing on the chiftop to the west of Portscatho at about 8. 00 p. m, when he leapt to his feet and exclaimed at the sight of an unfamiliar, large marine creature with a long neck, small head and large hump protruding high out of the water, with a long, muscular tail visible just below the surface, propelling itself in a north-northeasterly direction just offshore.

Drawing the attention of two passers-by with binoculars, we were able to scrutinise the grey, slightly mottled creature closely and observe that there was either another hump at the base of its spine, or more likely that the muscular rhythms of the tail created the appearance of a hump. The tail seemed to be about as long as the body and the creature was an estimated sev-

enteen to twenty feet in length.

For several minutes we were able to observe this graceful creature, with its head held proudly as it glided swiftly and smoothly on the glassy surface of the water, illuminated in the clear evening sunlight.

There were a number of birds wheeling around and a couple of boats in the vicinity, but all seemed oblivious of one another. Suddenly the creature submerged; it did not dive, but dropped vertically like a stone without leaving a ripple………" [13]

My ex-wife and I visited Shield Bird in the early spring of 1996 and we were both impressed by what a solid, sensible and down-to-earth woman she is. We spoke to her at length about her sighting and she told us again how she had agonised for months over whether to make her sighting public. Reading between the lines it seemed as if she was scared that she would be linked with all the media furore that had surrounded Tony Shiels's antics. Tony had become notorious across the country after the Morgawr sightings especially after his revelations that some of his magickal experiments were conducted by/with skyclad (naked) witches who happened to be his own daughters. His antics even made the national newspapers with *The Sun* describing the Shiels clan as "The Weirdest Family in the Land".

Twenty years after appearing all over the fortean and paranormal press in the guise of Psyche the skyclad monster invoking witch, Tony's daughter Kate wrote:

"Some serious feminists, deeply involved in the women's movement have been rather critical about allowing myself to be photographed nude in those days. They see it as a kind of exploitation, cheap cheesecake, or something of the sort. I disagree. As well as being a witch, I was in show business. I always retained full control during the photo sessions, and most of the photographers and reporters were a wee bit scared, afraid and in awe of the Shiels clan.

Nothing was ever published without my permission. If anyone was being exploited, you could say that the witches some of it anyway exploited the press. Yes, I know some people are shocked by nudity when it is associated with witchcraft. They see it as utterly wicked and depraved. I feel sorry for those small-minded puritans.

They must live horribly frustrated lives. I feel free to do whatever I wish to do, so long as it harms no one. I see nothing harmful in those photographs. For one thing, they help to dispel the popular notion of a witch as an evil, ugly old hag. That is just one aspect of witch nature, as perceived by man." [14]

I have only met one of the naked witches: Miranda only became involved in Tony's invocational activities relatively late in about 1980; but, in a quiet Cornish voice, she explained to me that witchcraft was women's magic and women's power. I am sure that she is right. Since I first became involved with Tony Shiels I have made friends with a number of people within the pagan community. I even share a house with one, and although I remain a Christian, I have nothing but respect and admiration for sincere neo-pagans who are only trying to live their lives in the way that they see fit.

However, I have a sneaking suspicion that Sheila Bird would not have seen it like that, and would have been horrified at the thought that anyone, especially the British media, would link her with such arcane goings on. I remain convinced that she is a sincere woman who saw something very strange in the sea that July day in 1986, and her testimony remains one of the best accounts on record of Morgawr the sea dragon.

It does seem that when one approaches the subject of lake and sea monsters that there are two very different types of phenomena at work. Some, like the creature reported by Sheila Bird or the animals killed or found dead at various points along the southern Cornish coastline are most certainly living creatures. Some may even be new species of animal unknown to science. Others are, as I have shown, most certainly nothing of the sort. My first visit to Loch Ness was in the spring of 1991 during a three-day break from a tour by *Steve Harley and Cockney Rebel,* a band for whom I worked at the time.

We drove right around the Loch, keeping our eyes firmly on the cold black water below us, convinced that we would see the elusive monster. As we drove past the gates of Boleskine House I felt a chill run down my spine. I realised that we were within figurative spitting distance of one of the most notorious sites (from the point of view of both a fortean and a Led Zeppelin fan) in the British Isles. Boleskine House was for fourteen years, was the home of Aleister Crowley, the self-styled 'Great Beast.' It has had several owners since Crowley, including Jimmy Page of the rock band Led Zeppelin. His daughter died tragically some years later and Page sold the house. Another owner tragically shot himself for no apparent reason. All this adds to the myth, and perhaps it is just as well that Boleskin House is not open to visitors.

I made the others stop the car and I got out to investigate. There were obviously people around, so I didn't dare to investigate too closely but I gazed in awe at the stone eagles on the pillars of the gate and looked longingly at the driveway, but I decided that as Jimmy Page (who still lived there at the time) was a notoriously reclusive bloke with a reputation for giving unwanted visitors short shrift, that I had better not investigate further.

The one thing that surprised me about Loch Ness was how open and exposed it was. Even in the early spring, however, every layby and parking space on the shores of the great peaty lake was crammed with carloads of tourists and wannabe cryptozoologists desperate to catch a glimpse of the monster. It was then that I realised that there would be a better chance of seeing a *bona fide* prehistoric survivor behind the bacon counter of your local *Sainsbury's.* However something most certainly lives there, and not all the sightings of anomalous animals in the lake can be explained by the appearance of out of place sturgeons, seals, or porpoises or the existence of outsized common eels, or even a hypothetical new species of anguilliform.

Something far stranger is in the *zeitgeist* of Loch Ness and, indeed the other so called monster-haunted lakes in Scotland.

Ted Holliday described, at some length a syndrome which besets monster hunters world wide:

"......... it was clear that the Morar monsters were acting in exactly the same curious way as the ones in Loch Ness. Either a camera was not available to record what was observed or, if it was available, circumstances frustrated the photographer. Almost everyone rejected such a notion because it introduced an element of irrationality. It also raised doubts about the true nature of dragons which those who were anxious to press the claim for an unknown animal chose not to encourage. Normal animals do not behave in such an inexplicable way because they cannot; therefore you had to conclude that the peculiarities were due to chance. This was the prevailing attitude amongst the investigators.

An explanation based on chance seemed to me most unsatisfactory. Chance is a random effect; it is just as likely to work in favour of the investigator as against him. If the ten years of intensive effort at Loch Ness which resulted in failure to get a detailed film was the result of chance then it was not a random effect and the expression became meaningless. In that event, the explanation lay elsewhere."

Over the next few years Holiday proceeded to collect a large number of *"examples of what appears to be some sort of a mental block in relation to the phenomena. There is a desire to minimise or dismiss what one has seen and this provides a brief interim in which the object escapes further observation."* [15]

This included testimony from such people as naturalist and author Gavin Maxwell, and led him to believe that this syndrome was somehow part of the nature of what he was studying. When he had come to this realisation it was but a short step to another series of revelations:

"Even more marked than the above were the situations in which monster phenomena seemed to be actively evasive. The Loch Ness Investigation Bureau's camp at Achnahannet, for example, has maintained a watch over this part of the loch for six months of the year every season since 1965. But no major sighting has taken place in this area since the Bureau set up its 35 mm. cameras and 36 inch lenses. Yet in 1964 there were two authentic views of monsters from this spot by multiple witnesses, which I have interviewed.

The apparent evasion of Bureau cameras by the phenomena is of long standing. An early case occurred in 1965 when a camera located on a platform near the Clansman Hotel was taken away for servicing. The next morning, the staff of the hotel had dramatic views of a hump moving about inshore. Over the years, quite a catalogue of such incidents built up!"

The twin pitfalls of psychic backlash (see Chapter Four) and this apparent inability to be captured on film have beset both monster hunters and UFOlogists for years.

One is left with a question that is fundamental to all monster hunters who are broad minded enough to accept that not all monster sightings are reports of hitherto unknown species of animal. Are places strange because they have monsters in them? Or are monsters attracted to strange places?

Evidence from another famous monster haunted lake suggests the latter. Consider this USO (Unidentified Submarine Object) from Kelowna, British Columbia.

"It was a clear bright morning in September, and when we reached the ferry slip to cross over to Kelowna we realised we would have to wait a bit because we could see the ferry still on the other side. And then about half a mile north of the ferry we noticed this other thing. I remember pointing it out and saying what a beautiful white boat it was. It was moving around so gracefully, and though we still couldn't make it out very well at that distance, it seemed to have a smooth round design we had never seen before."

As the three watched from their car in admiring curiosity, they noticed the strange craft had started to move across the lake in their direction. By that time cars for the ferry were beginning to line up behind them, so they were also in a position to observe what happened next.

"We could see the wash coming out from either side, yet somehow the boat, as we thought it was, didn't seem to be moving very fast, " Stewart said. (Dorothy compared it to a line from Dante, "Hasten slowly.") "As it came closer we still thought it must be some unusual kind of modern boat. It looked like a round hard hat sitting on a platter. But there was something about that was that looked different and that started us wondering."

Although neither could explain precisely what the difference was, possibly it was caused by the circular shape of the craft moving lightly on the surface, like a flatly thrown stone. "Then suddenly it really surprised us, " Stewart continued. "It was a few hundred yards away when all at once the wake disappeared and we realised the thing was in the air. It changed direction to the right so that it came straight toward the ferry dock and then it stopped dead, less than 100 feet in front of us and about 50 feet above the water." [16]

It is interesting that their experience should have happened in that particular location for *Lake Okanagan* is perhaps better known as the home of another well known quasi-fortean phenomenon, a lake `Monster` known by the charming, if slightly annoying appellation of `Ogopogo`.

Ogopogo is not an Indian name for the world-famous, friendliest inland sea monster. The name is derived from a music hall song that was popular in the 1920's.

The Ogopogo Funny Foxtrot

I`m searching for the Ogopogo
the funny little Ogopogo
his mother was an earwig
his father was a whale
I`m gonna put a little bit of salt on his tail
I`m searching for the Ogopogo
as he`s playing on his old Banjo.

*The Lord Mayor of London
The Lord Mayor of London
The Lord Mayor of London's
gonna put him in the Lord Mayor's show* [17]

Interestingly enough, despite claims to the contrary in books like *In Search of Lake Monsters* by Peter Costello, the animal described in the song and pictured on the front cover of the sheet music isn't at all reminiscent of the animal reported from the lake. Rather than being a water dwelling leviathan, it is a small humanoid creature with antennae like those of a bumble bee.

The main protagonist of the song is *"a funny little man"* from *"the hills of Hindustan"* who wears plus fours and whose modus operandi whilst chasing the elusive Ogopogo is to use an Elephant Gun as well as the tried and tested method of putting *"a little salt on his tail"*.

In light of this evidence it is tempting to theorise that the appellation was granted to the lake monster because the sophisticated settlers of European origin DIDN'T believe in the monster, and felt that it was as amorphic and ridiculous a creature as the one described in the song, rather than because of any supposed physical resemblance between the two beasties.

Indians referred to Ogopogo as N'HA-A-ITK which when translated means "Lake Demon". Legend explains that the creature was actually a demon-possessed man who had murdered a local known as Old Kan-He-K. (Lake Okanagan was named in his honour). As punishment, the native gods turned the murderer into the giant sea serpent so he would remain at the scene of the crime for all eternity. Hence Ogopogo's longevity. To appease the monster N'HA-A-ITK (Ogopogo), the Indians offered small animals at its legendary lair/submarine caves off Squally Point near Rattlesnake Island. Ogopogo frequents the waters between his favourite island and Mission Valley and has made journeys to both ends of the lake. Recorded sightings date as far back as the early 1800's. In 1860, John McDougal lost his team of horses when they were pulled under as he was swimming them across the lake in a canoe . . . . never to be seen again. [18]

Ogopogo is dark green in colour, estimated at one to two feet in diameter with a length ranging between 15 to 50 feet. Ogopogo's head is said to resemble that of a horse or goat head with a beard. Ogopogo has been mistaken for a log, boat wake, large sturgeon and other floating mysteries.

The government, in 1926 announced that the new ferry being built for travel across the Okanagan Lake would also be equipped with special "monster repelling devices". Since the construction of the floating bridge, it is assumed that the bridge has enough support and strength to withstand any nuzzling or advances of Ogopogo. Travellers' safety while crossing the floating bridge is assured as maintenance crews are often checking for and repairing any damage.

Ogopogo has been sighted by many individuals who have remained firm in their belief, despite the ridicule from legions of nonbelievers. Both sides in debate seem to be divided into equal camps. The majority of sightings have been consistently similar. The "fearsome thing" is generally described as having a snake-like body about 20 feet long, dark green skin, with a bearded horse or goat shaped head.

The correlation between monster sightings and other fortean phenomena is one that occurs again and again. The files of the CFZ are full of such cases, and some of them are included in this book. What exactly that link is, however, remains obscure although one thing is certain. Although some of the animals that I have hunted and which are described in this book are *bona fide* unknown species of animals, many of the strangest are nothing of the kind and so long as investigators insist on treating creatures like Morgawr as if they were just another species of ordinary animal, albeit one whose existence is presently unknown to science, then the mystery will never be completely solved!

However, back where we started on Maenporth beach in 1998 the TV Crew, Graham and I stared at the sea for several minutes, but absolutely nothing happened at all! Toby cocked his leg and peed all over one of the ritual candles and Graham mistaking one of the other candles for a piece of flotsam washed up on the beach drop kicked it into the sea in a manner reminiscent of that football playing geezer who married one of The Spice Gurrls.

"Oi" shouted Richard in his normal voice *"you've broken the bloody circle. The invocation will never work now!"* Everybody laughed (except Richard who was mildly put out as he had hoped that the ritual would be a success and we all adjoined to the boozer in Mawnan Smith for lunch.

Richard's fascination with dragons is well known. He led an expedition to northern Thailand in October 2000 in search of a semi legendary Thai dragon called the Naga but he is also a ritual magician with a deep-rooted interest in dragons. He had been planning his own serpent raising ritual. He was approached in February 1998 to just such a conjuration for a local T. V company who was filming a documentary on sea serpents. Resplendent in ritual robes and wielding his ritual sword he obliged. Unfortunately high magick and T. V companies do not make the best bedfellows. The ritual was halted, restarted, held up and re-shot from different angles. Cries of *". . . . . . . can you just evoke those dragons of the western quarter one more time for sound"* and such became all too familiar. In short the ritual was never really carried out, it was just a set piece for the cameras. But it seemed that like Tony Shiels so many years before, his shenanigans actually *did* have some affect . . .

In August 1998, two 15-year-old Topsham girls, Laura Malkin and Hayley Wannell, were walking along Saunton sands when they noticed an odd shape in the water. As it drew closer they could make out what appeared to be a snakelike head and neck and a large humped body. With shaking hands they pulled out a pocket camera and took a snap of the object. Within days the Exeter Strange Phenomena group was called to the offices of the Express and Echo - the local newspaper that was about to run the story. They wanted a trained zoologist to examine the picture before they went to press.

Wearing the guise of an experienced zoologist rather than a magician, Richard Freeman took a close look at the monster. The creature did indeed bare a resemblance to the classic picture of a sea serpent but something worried him. The animal's outline seemed very irregular and after consideration he was forced to dismiss it as a hoax. A couple of days later he was vindicated, the sea monster turned out to be nothing more esoteric than an oddly shaped log.

A fake monster raising ritual seemed to have spawned forth a fake monster!

REFERENCES

1. The `Hag` aspect of the Goddess
2. *Falmouth Packet* 2. 4. 76
3. 1876 report in *The West Briton* cited by Bright, M. , *There are Giants in the Sea* (Robson, London, 1989)
4. Bright *op cit*
5. Heuvelmans B *In the Wake of the Sea Serpents* (London, Hart-Davis, 1968)
6. Wilkins, H
7. Mawnan-Peller (Pseud.) *Morgawr – the monster of Falmouth Bay* (1976)
8. Issue 8
9. Geller U *The Geller effect*

The Dobhar Chu

Ireland's not so mythical, occasionally lethal and currently enigmatic otter

by Gary Cunningham

Legend, folklore and early accounts.

The island of Ireland is a country very rich in legends of long since departed heroes, kings and warriors and the epic battles, which were fought, monsters slain or tales of bravery and skill

Names such as Cuchulain, Oisin and Finn Mac Cumhail are integral characters in such stories. They were also said to have encountered fearsome monsters in Ireland's waterways. There are, of course, modern day reports of lake (or lough as they are known in the country) monsters. The most well known, mysterious inhabitants of the loughs, locally termed 'horse-eels' have been seen quite regularly in particularly in the 1950s, '60s and '70s from the western portion of the island and more specifically from the sparsely populated area called Connemara in Co. Galway.

There is, however, a very different creature, which according to Irish legend, did not share the same physical features of the horse-eels. This creature was known as the Dobhar-Chu (pronounced Dovar-Ku) and was said to have been a huge breed of otter with supernatural qualities! Within Ireland, this supposedly 'mythical' beast has vanished from memory with one notable exception - the County of Leitrim bordering counties Sligo and Donegal. The local population has very good reason for being unable to erase this creature from their memories and folklore - one of their own was said to have been killed by a Dobhar-Chu.

The Dobhar-Chu was believed to inhabit a lake known as Glenade Lough. This lough is lo-

cated in the southern portion of the county and can be reached by following the R280 northwards from Manor Hamilton to Bundoran. The lough itself is approximately a mile in length and a half-mile in width, and is on the left side of the road for at least some of the road's progression northwards.

Local legend tells of a woman by the name of Grace Connolly, although some versions call her Grace McLoughlin or McGlone. In fact, this is partly true, as her married name would have been McLoughlin, only the custom was for the woman to retain her maiden name. Grace is said to have gone to the lough to wash her clothes (again some versions have her bathing in the lake).

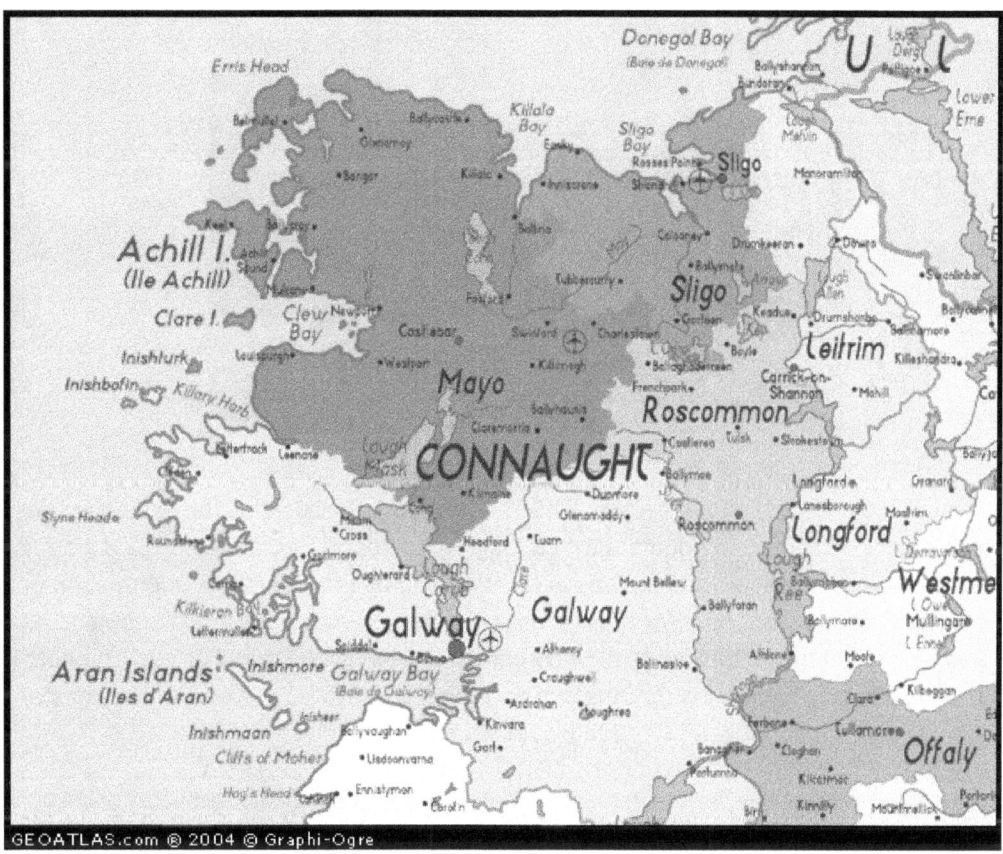

When Grace failed to return, her husband Terence went to search for her. When he eventually found her she was fatally wounded (and had been presumably mauled). The sight that confronted him must have been terrible for, lying across his wife's motionless body was her assailant, a Dobhar-Chu. Terence killed the beast. However it's dying cries were sufficient to attract its mate from the lough. This second Dobhar-Chu relentlessly pursued Terence (who was now accompanied by his brother - Gilmartin). Both men fled on horseback.

The two men eventually reached a place, about 20 miles away, known as Castle Garden Hill. They then placed their horses in such a manner to barricade themselves from the Dobhar-Chu whilst the two men lay hidden waiting for the beast to show. The 20 miles or so of intervening mountains and difficult terrain posed no problem for the Dobhar-Chu as it eventually reached where the two men lay in ambush.

In some versions of the story what proceeded next must have been savage as the Dobhar-Chu killed one of the unfortunate horses and before the men's remaining steed suffered the same fate, Terence surprised Dobhar-Chu stabbing it fatally.

Glenade Lough, Co Leitrim, Ireland: This relatively small lough was the setting for one of the more mystifying encounters with a lough monster. In September 1722 a woman by the name of Grace Connolly was allegedly killed by the legendary Dobhar-Chu or Master Otter - a giant otter (or at least otter-like beast) whose existence seems to be well testified in this otherwise serene corner of North-West Ireland. NB View in photograph is looking west towards Drumcliff and Sligo Bay. Note the weed filled edges of this picturesque lake.

The event is also commemorated in verse, as there are at least two poems dealing with the unfortunate death of Grace Connolly by her loathsome assailant - the Dobhar-Chu. One of these poems was called "The Old House", and was included in a 1950s collection - *Further poems by Katherine A Fox* who was herself from County Leitrim. The lines, which mention the Dobhar-Chu, are as follows:

"The story told of the Dobhar-Chu,
that out from Glenade Lake
Had come one morning years ago

A woman's life to take."

The above lines are quite basic but demonstrate that the event was important (and possibly devastating) enough to have been recorded literally in order that it's memory should not be forgotten. The second poem is written anonymously and is simply entitled – "The Dobhar-Chu of Glenade". It goes into greater detail than the first and, in its entirety, totals 16 verses and is therefore extremely valuable to the whole legend.

Photograph showing the complete tombstone of Grace Connolly with the Dobhar-Chu at the top of it.

The tombstone is horizontally laid in the cemetery and is approximately four and a half feet (135cm) by two feet (60cm). It is made of sandstone and can be located within the cemetery by walking roughly twenty feet (6.7m) in a north-westerly direction after braving the small path from the entrance gate.

The legendary Dobhar-Chu (Master Otter) depicted on Grace Connolly's gravestone in the town land of Drummans, Co. Leitrim, Ireland. The animal depicted on Grace's tombstone is the very one said to have killed her whilst she washed her clothes at Glenade Lough (some 3 miles distant from the cemetery). It is very striking and contains features of an otter, such as the tiny ears, long sturdy neck and large paws but also possesses characteristics of a dog with long powerful limbs, a deep barrelled chest and lengthy tail (which apparently had a tuft at its tip although this feature is now difficult to discern as a thin sliver of the sandstone of which the tombstone was made has sadly broken off. What is this strikingly unorthodox animal's zoological identity?

The date on the tombstone is very hard to discern but is dated September 24 1722. As can be seen from the photograph much of the writing has eroded due to rain and time, and the majority of the tombstone itself is covered with lichens (the large splotches of white and grey). This animal's strange identikit is as follows:

- Its hind legs and front legs are equally long and powerful.
- It has a long tail with a barely conspicuous tuft at its tip (the portion showing the tuft was a thin slice of sandstone and has since flaked off)
- It also has a deep set powerful chest.
- The creature's small head and paws are very like those of an otter.
- Its neck (which is doubled over it's shoulder and consequently can be missed when first observing the tombstone) is also reminiscent of an otter.
- It's barrel-like chest and long muzzle are morphologically canine.

One question that I believe is fundamental to the entire mystery from a historic sense is, if this tragic incident did not occur then why portray the creature which attacked and killed Grace Connolly upon her gravestone in such detail?

By comparing dates and styles of other old gravestones in the cemetery, we can determine that Grace did die at around the once visible date on her gravestone - September 1722. Interestingly, other gravestones, which seem to be from the same period as hers, have the surname McLoughlan on them.

This is the variant of Grace Connolly's husband's, name (Terence McLoughlin) and is stated as such in some versions of the tale. The graveyard, as previously mentioned, is situated on the right side of the R280 to Bundoran from Manor Hamilton and approximately seven miles from the Leitrim village. The cemetery is noted as an antiquity (ancient monument of interest) on the Ordnance Survey map of the region (no 16 in the Discovery series) and is located in the town land of Drummans. A nearly complete signpost still indicates its position from the side of the road.

I visited the Conwell (or Teampall as can be seen in the photograph) cemetery in July 2000 in order to locate the Dobhar-Chu gravestone. After about 20 minutes searching I found it. It was positioned horizontally in the ground not too distant from the gated entrance. I was excited at actually locating it as I had searched in vain for my quarry in April 1999 in horrendous weather - there was torrential rain at the time!

Sign still indicating the graveyard and ancient church at Conwell cemetery and most significantly the Dobhar-Chu tombstone. (*To*.... Is all that remains of `tombstone` on the sign). This sadly incomplete signpost is located on the left side of the main Manor Hamilton to Bundoran road (R280) and is directly opposite the gated entrance to the graveyard. The site is approximately seven miles on this road from Manor Hamilton.

I took several photographs (which are included in this paper) of both graveyard and tombstone and also took some video footage of the surrounding environs. Also intriguing is that the barbed spear grasped by a gloved hand (as can be seen from the enlargement of the photograph of the animal) enters at the base of its neck above it's powerful chest and re-emerges below where its ribcage/stomach would be. This is thought-provoking because if the depicted animal is one of fabricated myth and legend then why would the stonemason have gone to considerable effort to closely tally one of the most significant strands of the alleged event? i.e. the slaying of the Dobhar-Chu by the avenging man - Terence McLoughlin.

Up until the First World War a second Dobhar-Chu tombstone existed not too far distant from the first at a town land called Kilroosk. This place is to be found approximately three miles north of Manor Hamilton and can be reached by taking a third class road from the adjoining second class R280 (as already mentioned). Interestingly - on the Ordnance Survey Map No 16 in the Discovery series a red dot denoting a named antiquity is visible along with the letters Ch (which must stand for church).

This second tombstone was also said to have shown a Dobhar-Chu, although when investigated by Patrick Tohall (who had written a very informative article for The Royal Society of Antiquities in Ireland sometime after 1935), the local men referred to it as the Dobhar-Chu stone and the only man who remembered it's image told of a creature like a horse!

It was the gravestone of Grace's husband Terence (or Ter as is written on the tombstone of Grace) McLoughlin. However, sadly at one stage this stone was placed on a boundary wall after it had broken in two and was subsequently lost! The very fact that at one point in time two gravestones existed within the same geographical area demonstrates how significant the event was to the local population.

Indeed, for both gravestones to depict the animal which instigated this tragic occurrence is indicative of how unusual such a creature was and possibly so, too, the death of Grace McLoughlin!

Also worthy here is the mention in Patrick Tohall's article that the incident was still fresh in local memory thereby fortifying the entire occurrence as fact, and also the appearance of Grace's assailant. It is very relevant to summarise the main strands of the legendary and mythical qualities attributed to this enigmatic beast.

- Firstly it was believed to have been of an unusually large size sometimes holding court with five or six regular sized otters.
- Secondly (and possibly significant when endeavouring to identify it with a known group of animals) its mythical qualities included the notion that even the slightest portion of its pelt could save a ship from being wrecked, a horse from drowning or even a man from gunshot or any other serious affliction.
- Thirdly it was also the belief that the animal itself could only be killed by shooting it with a silver bullet (this bizarre notion is uncannily familiar from the werewolf legend).
- Finally, Patrick Tohall, whilst researching the Dobhar-Chu, was informed by an old man

in the records of the Commission for Boundaries, Co. Donegal, that an old Irish phrase stated that the Dobhar Chu is the seventh cub of the common otter.

In fact, the very usage of the word Dobhar-Chu for the Irish subspecies of Eurasian otter is now replaced by Mada Uisge, and both phrases literally translate as 'hound of the flowing waters' or simply 'water-hound'.

The once oft seen, but never captured mysterious inhabitant of Sraheen's Lough.

When we tend to think of Irish lake monsters we picture images and stories of those equine-headed eel or serpent-bodied creatures aptly named Horse-eels. These still enigmatic inhabitants of Irish lakes (loughs) have been most frequently seen in the sparsely populated region of Connemara, Co Galway. The overall morphological identikit of these sinuous animals would seem to fit very well with the artistic reconstructed appearance of the supposedly 'extinct' sub-order of early whales - the Archaeocetes (or ancient whales). To give these unidentified lake monsters proper coverage would involve dealing with them separately in a future article.

Sraheen's Lough, Achill Island, Co Mayo, Ireland: The lough is almost surrounded by thick rhododendron bushes and offers sanctuary for any unknown animal or animals whether they are transient visitors or were indigenous to it and its environs.

However some denizens of Irish loughs cannot be categorised in this manner and one of these (possibly the last of its kind) was sighted in a small body of water on Ireland's largest island - Achill Island in Co Mayo. This lough is called Sraheen's Lough (although it also goes by the name of Glendarry Lough). The lough is overtly circular in shape and approximately 400ft

(130m) in diameter. When surveyed by Australian students in the late 1960's it was deemed bottomless - possibly the now filled crater of an extinct volcano.

There are dense rhododendron bushes growing around most of its shoreline and considering it's location on one of the most westerly regions of land in Europe, a more windswept, weather-beaten isolated wilderness devoid of extensive forestation would be difficult to conceive.

Achill Island, as noted earlier, is Ireland's largest island some twenty miles long (32km) by fourteen miles in width (23km). It possesses mountainous peaks some two thousand feet plus (620m) and also the highest sea-cliffs in Europe where the northern face of Croaghaun mountain drops nearly 2200ft (690m) into the Atlantic below, not too distant from the most westerly point of the island itself at Achill head. Also very significant within the context of the legend and possible existence (at one time) of the Dobhar-Chu is its geographical location and, indeed, proximity to Glenade Lough in County Leitrim.

Achill Island is situated just off County Mayo, which is the county immediately to the west of Counties Sligo, and Leitrim, where the Dobhar-Chu legend to still an integral part of daily life. There was an oral tradition of water monsters in and around Sraheen's Lough on the island. Stories and tales were remembered by the older generation (and used to frighten unruly children) but in 1968 such stories were to become fortified by an unbelievable series of events, thus allowing fairytales to make the transition from myth into reality.

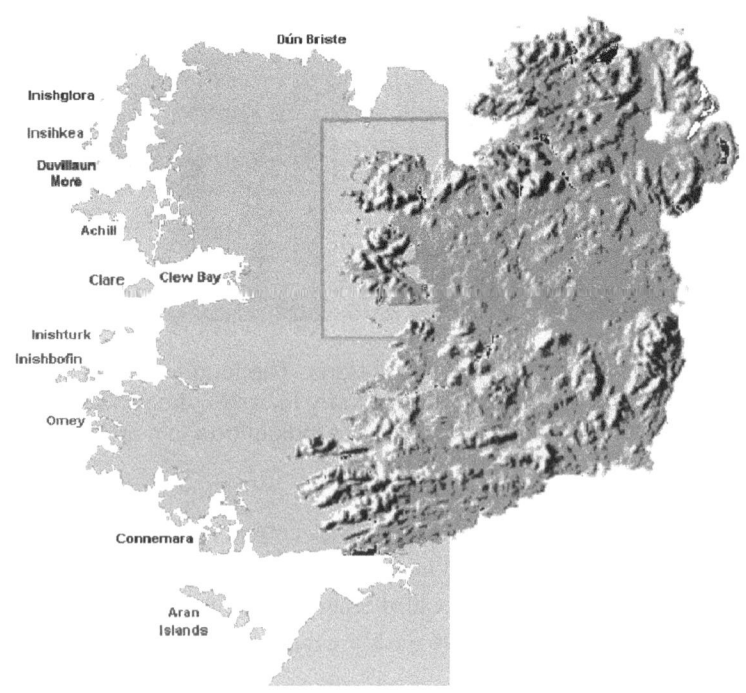

It all began on the 1st May 1968 when two local men - John Cooney and Michael Nulty were to have an unforgettable experience of a strange creature at the lough at night. They were driving home to Achill Sound on the mainland from the village of Keel when, at about 10:00pm, a bizarre animal ran across the road from the direction of the lake and disappeared into the thick undergrowth. They were able to see it clearly as it was illuminated by their headlights. They described it as being eight to ten feet long (2.4 to 3m) four-legged, black or dark brown in colour, having a thick tail, a long neck like a swan and a head like a sheep or a greyhound.

It was also about two and a half feet tall (0.75m) and when it ran it rocked from side to side. John Cooney also noted that it was moving at an angle, weaving and curving (which suggests that the animal was moving in a haphazard or random way). Mr. Cooney also mentioned that its eyes were 'glittering'. This would suggest that the iris of this animal were highly reflective and might indicate that it belonged to a mammalian species.

Due to the nature of their incredible and unbelievable sighting the men certainly didn't stop to investigate further. The men's sighting, however unusual was not an isolated one and only a couple of months later two girls were hitching a lift whilst on their way to the mainland, when they too thought that their eyes were deceiving them.

Mary O'Neill, who was from Mullingar, Co.Meath, and Florence Connaire from Galway, were given a lift by a Dundalk businessman (whose identity I have not been able to discern) in June 1968. Just as they passed the lake one of them noticed a huge animal on the lakeshore approximately 100 yards (95m) from their position. The animal was about 20 feet in length (6.5m) (although the size is doubtful and obviously exaggerated).

It also possessed a head like a greyhound's (sounds familiar!) and a long tail.

This sighting gained notoriety in the Irish press as apparently the businessman stopped and took a photograph of the animal. An enlargement of the photograph, accompanied by an account of the incident, was published in the *Dublin Evening Herald* on June 5, 1968.

However, this sighting was probably an invention by someone eager to try and 'cash-in' and exploit the foundation of earlier sightings. As the creature in the photograph was said to have resembled a dinosaur, and due to the anonymity of the Dundalk businessman involved, the photographic 'evidence' obtained is now believed to have been a hoax.

Other sightings (and in particular two detailed ones), were to occur in and around Sraheen's Lough, which were equally as incredible as that of John Cooney and Michael Nulty. Two years before their sighting, in July 1966, an Englishman was fishing on the lough. Suddenly the serenity of the lake was disrupted by a commotion on the surface. The fisherman watched in awe as a long neck possessing a swanlike head appeared. The long neck was connected to a shiny black body, which also soon emerged. The creature then decided to move towards the petrified angler. He then fled to a shop on Achill Sound - approximately a mile away – which was owned by a man called Dennis McGowan. This is where, apparently white with fright, he recounted his incredible experience.

Only a week after Messrs. Cooney and Nulty's sighting, a fifteen-year-old boy, Gay Dever, stopped whilst cycling by the lough on his way home from Mass. He had dismounted from his bike and was pushing it up the slight gradient by the lough when he heard splashing sounds. It was early evening and what he saw was a creature emerging from the lough and climbing a bank.

Why would an artist's reconstruction of a creature seen on Achill Island as recently as 1968, and the creature depicted on a tombstone nearly 280 years old of an animal of Irish legend - the Dobhar-Chu (which had by all accounts savaged an actual person - i.e. Grace Connolly) be strikingly similar in body shape and design? The answer (incredulous and improbable) would seem to be that the creature of Irish legend - the Dobhar-Chu or Master Otter, was very much extant (living) and not extinct and had been residing at a small lough on one of the most isolated regions of Europe as recently as 1968

Visibility was still good as there was adequate light and he watched an animal 'bigger than a horse' (a possible exaggeration initiated by such a shocking unprecedented sight) with a sheep-like head on a long neck, with a long tail and possessing four legs. The hind legs were much larger than the front ones and moved in an unusual fashion.

It moved in a rather jumpy way "like a kangaroo" and was about 12 feet (4m) long. As already mentioned there was a strong oral tradition of a 'monster' in the lough with many sightings claimed (particularly in the 1930s). The incidents on Achill Island may seem at first glance to be unrelated to the legend of the Dobhar-Chu. However there is a very fascinating link between the two. As highlighted and brought to the attention of the cryptozoological community by Dr Karl Shaker in an excellent article on the Dobhar-Chu in *Strange Magazine* #16 for his

"Menagerie of Mystery" column.

An artist's reconstruction appeared in an article on Irish Lake monsters by Janet and Colin Bord in a book entitled - *Creatures from Elsewhere*. This book was published in 1984 and is a compilation of articles featured in *The Unexplained* - an early 1980's part work.

The artist's reconstruction, based on the eyewitness accounts of 1968, is strikingly similar to the animal depicted on Grace Connolly's tombstone in Drummans, Co. Leitrim. Both creatures possess the same powerful chest, long tail, muscular legs, long neck and a dog like (described by eyewitnesses as greyhound like) small head with pointed ears. There are large paws on both front and hind limbs.

The artist's reconstruction of the Sraheen's Lough monster is eerily similar to the animal depicted on the gravestone of Grace Connolly in Co. Leitrim. Both creatures have been separated by over 240 years. Is it possible that a specimen of the Dobhar-Chu was alive and well and living on Achill Island as recently as 1968?

Its mythical qualities offer up clues and possible solutions.

The identity of the animal responsible for the legendary Dobhar-Chu is not as lucid as it would first appear; indeed the very subject seems complex, with various strands entwined in the myth. However certain characteristics of the Dubhar-Chu and what was known to the local population may enable its identity to become more apparent when we look at them objectively and from a zoological perspective.

- First and foremost an intriguing, even supernatural aspect of the creature's legend is the magical quality of its pelt.

According to the legend - 'an inch of its coat will prevent a ship from being wrecked, a horse from injury and a man from being wounded even by gunshot'.

I believe that this strand of the legend can be linked to a fascinating fact concerning another member of the Mustelid family - the sea otter (*Enhydra lutris*). This endearingly cute mammal has the distinction of possessing the thickest coat of all the Mustelids (the group including otters, weasels and badgers). It has the highest density of hairs (126,000 per square centimetre on average) and there is a very important reason for this.

Due to the sea otter's almost totally exclusive existence in the cold, harsh environment of the Pacific Ocean (it doesn't even come ashore to breed and even gives birth at sea) it needs to retain body heat as it lacks the insulating layer of fat known as blubber which is possessed by larger marine mammals such as seals and whales. Its luxurious pelt was the reason for the sea offer's near demise in the 18th Century. Fur trappers killed some 190,000 plus otters in order to meet the demand from the then non-conservation minded population of the USSR.

Here, after the sea otter's discovery by a Russian expedition to Alaska in 1741 and transporting their fur back to their country, the value of its coat as a luxurious commodity even exceeded that of another Mustelid - the sable. The primary function of the sea otter's coat is to trap air due to the sheer quantity of hairs. The layer of trapped air acts like an insulating material/layer for the animal. Consequently this incredible physical attribute of the sea otter may have inspired the beliefs about the Dobhar-Chu. Remember that according to legend – "even an inch of its pelt is enough to save a man from injury".

Its unique coat would indeed seem impervious to any attempts to penetrate its richness and thickness. Also very notable with regard to the physical appearance of the Dobhar-Chu, is the belief that according to legend, its "white coat" was said to have a black cross on the back of the creature.

This very strange feature was not the only unusual variation with regards to the creature's pelt colour as the tips of its ears and the end of its tufted tail were also said to have been black. I wonder if the legend is (and was) highlighting another feature of the animal's appearance, which could link it to an animal group. Certain members of the Mustelids and more specifically the stoat (*Mustela Erminea*) develop a white coat in the winter months as an adaptation to the changed environment it finds itself in.

However, such a change by the stoat is not complete, as the tip of its tail, the end of its ears, and its paws remain a dark brown in colour. Sound familiar?

This change aids the stoat in camouflage when being hunted by predators such as buzzards and eagles (to blend in with the white background of snow). The only major problem by endeavouring to reconcile the Dobhar-Chu with an animal reminiscent of or even related to the sea otter is that the latter animal is only found in the Pacific Ocean along certain coastlines, and not in the Atlantic. However, if one day a giant relative of the Pacific sea otter is discovered from the fossil record, and furthermore, one that resided in the Atlantic Ocean, then we would be presented with a notable precedent for a real animal identity behind the 'mask' of the Master-Otter or Dobhar-Chu. In fact there *is* a fossil otter, which in life was adapted to a marine environment.

This species was known as *Potamotherium*. It lived during the Oligocene period and early Miocene (some 38 to 25 million years ago) and its fossil remains have been found in Europe, particularly so in parts of France. From the evidence presented in this chapter (regarding the animal's coat) I believe it is safe to assume that whatever animal was responsible for the Dobhar-Chu of legend and fact, it must have had more than a transient relationship with the intriguing family of mammals known as the Mustelids.

Proposed identities for the Master-Otter.

Indeed if we consider identities for the Dobhar-Chu/Master-Otter we cannot rule out the possibility it may have been a surviving member of a prehistoric mammal lineage.

The megafauna of the last Ice Age may have included the direct ancestor or species of the legendary beast in the form of a giant otter (marine or freshwater). It may also have been totally unrelated to the Mustelids and a possible identity could have been an animal known as a Condylarth. This unusual early group of mammals evolved into meat-eating mammals but still managed to cling on to their herbivorous ancestry by possessing physical characteristics more suited to a vegetarian lifestyle such as molar teeth and hoofed feet!

In fact some 50 million years ago these "wolf-sheep", as they are sometimes known, gave rise to the ocean-going whales and are relevant to the identity of the Dobhar-Chu. The animal from this group, which first ventured into an inland sea or lagoon in search of food most probably, resembled a large otter. This animal is called *Hapalodectes* and if its immediate descendants grew larger than it did (it attained 4ft) say 8ft in length, then we would have a very close match for the Dobhar-Chu if it survived relatively unchanged.

However more than likely with a time frame of 50 million years of evolution to the present day morphologically this would result in the animal looking very different from its current fossil and reconstructed appearance. A more convincing candidate from the fossil record is the aforementioned marine form of otter, which should be discussed in more detail.

Figure 24. The Oligocene otter *Potamotherium*, a very advanced water-living form which may be close to the ancestry of the fur seals. Fossils are especially plentiful in parts of France.

Age of Mammals
Bjorn Kurten

The prehistoric otter - *Potamotherium* - was more highly adapted to swimming than the present day otter - *Lutra*. It was better adapted due to certain physical characteristics such as more powerful limbs and larger, paddle like feet. Indeed as mentioned by world-renowned palaeontologist - Bjorn Kurten, *Potamotherium* possessed many seal like characteristics. From the artists reconstruction above (taken from Kurten's book - *The age of Mammals*) we can also see that this ancient form had small external ears (similar to the present day fur-seals), a long neck region and a sleeker head profile together with a slightly longer muzzle. These features along with the aforementioned longer, powerful limbs - particularly the hind ones) and the paddle-like feet do indeed give the animal a seal-like appearance.

Worth mentioning here, I feel, is the Mustelid's tendency towards specific physical attributes, which would assist a great deal in detangling the mystery of the Master-Otter. According to Sean O` h-Eochaidh (Sean Haughey) from Donegal (as mentioned previously) the Master-Otter was said to have been the seventh cub of a seventh cub and therefore a super otter, presumably a giant.

The otter family's tendency towards gigantism is evident in the fascinating Brazilian giant river otter or saro (*Pterona Brasilienis*). This highly social animal is not only special due to its gregarious behaviour (it is often seen in large groups consisting of many family members) but also attains lengths on average of 6ft (1.8m) or, in some instances, even longer - 8ft (2.4m) (however with over hunting and poaching such sizes are nowadays exceptional).

Sadly, this impressive animal's distribution is under threat from poachers and ironically tourism itself as this irreversibly affects the habits of this beautiful creature in its natural environment. Not only is gigantism evident in otters, but also the large body size attained by the sea otter demonstrates that otters have evolved quite comfortably with greater body size than most mustelids (it can reach the weight of 100bs (45kgs) and this is because of its requirement of living in the open ocean of the Pacific. By being relatively large the sea otter loses less body heat as it has a larger surface area. It can also attain lengths of up to 5ft 9in (1.6m) and so is by no means small and as long as the common European otter (*Lutra lutra*) although more heavily built.

Other identities for the Dobhar-Chu include a giant species or subspecies related to the European otter. This species or subspecies I believe may have been indigenous to Ireland with strongholds in the western region of the country. Primarily the counties of Leitrim, Sligo and Mayo, but also including Donegal and Roscommon to the list - based on possible sightings - perhaps as far apart chronologically as the late 1800's and early 1960's.

A genetic bottleneck caused by factors such as inbreeding, geographical barriers (such as mountains or islands) or even recessive genes from its evolutionary history could have resulted in a radically different otter. Such an otter would possess certain morphological and behavioural characteristics and traits not normally observed in the common Irish otter. Such unusual behaviour and traits might include:

- gigantism
- aggressiveness
- a greater carnivorous persuasion
- pure melanism

One further point which demonstrates that even modern day species of otters can exhibit unobserved and unprecedented behaviour can be seen from the following (possibly unique) example which parallels part of the Dobhar-Chu legend.

A pair of Brazilian giant river otters kept at Sao Paolo Zoo in Brazil both took part in killing a keeper whom they perceived as a potential threat to their young. I have not been able to determine the date of this tragic occurrence, however, and it should certainly be kept in mind the next time an over exuberant tourist with a camera wishes to photograph this beautiful, yet occasionally deadly, animal

Both the traits of gigantism and aggressiveness (and possibly a greater carnivorous tendency towards prey species) are/have been exhibited by the Brazilian giant otter. This relates quite well with what is known (or was known 'first-hand' in some instances) with both morphology and behaviour of the Master-Otter or Dobhar-Chu.

The Scottish 'angle'

When we consider the Dobhar-Chu legend, and its transition from folklore into fact, we must also examine its existence within not only Irish mythology, but also that of our northeastern neighbour - Scotland. A reference to the Dobhar-Chu in Scotland was kindly sourced and forwarded to me by fellow lake monster researcher, Nick Sucik, who is from Minnesota, USA. Nick is very interested in the Irish phenomenon and is well versed with regard to it. I am very grateful to him for the following reference.

White faced otter - called by the Irish Dobhar-Chu:

Martin in his interesting description of the Western Islands of Scotland, London 1703, vol 8, p159, tell us, that in the Isle of Skye, the hunters say there is a big otter above the ordinary size, with a white spot on its breast, and this they call the king of otters; it is rarely seen, and very hard to be killed. Seamen ascribe great virtue to the skin, for they say that it is fortunate in battle and that victory is always on its side.

Notable in the above account is the location where it can be found. Not only does this indicate to us a sense of former range and distribution but possibly also a preference for a particular type of habitat in this instance islands. The Isle of Skye and Achill Island would seem to have been populated or visited by the Dobhar-Chu.

An examination of vegetation type, topography and flora/fauna present and surrounding marine ecosystem might prove useful in determining why this animal chose these regions and so too its scarcity/population status.

As the Dobhar-Chu is included in Scotland's mythology then there may be at best be a link and at the very least a tantalising hint of the animal's geographical distribution at the present (or at least in the past anyway!). The animal may have resided in the furthest and least explored reaches of the Artic Circle in regions such as the coasts of Scandinavia or further west such as Iceland or even Greenland.

Its occurrence within the two countries mythologies may indicate that it was a visitor during its migrational route or maybe individuals being swept off course southward similar to immature walruses occurring along the coasts of Ireland in the last 100 years or so (incidentally this may be the identity of two very unusual creatures allegedly seen in two unrelated lakes in Counties Gaiway and Monaghan during the 1960's).

Scotland is also closer in proximity to regions such as the lower reaches of the Artic Circle, Scandinavia and even the Faroe Islands and the very basic fact that the creature is to be found in Scottish Mythology certainly pertains to the idea that the Dobhar-Chu range is located in this part of the world.

With reference to an identity for the Dobhar-Chu I feel that the fossil species of otter *Potamotherium* would be an excellent candidate for two main reasons.

Firstly (as can be seen from the reconstruction) it has a sloped profile mainly due to its longer hind limbs and coupled with the broad, paddle-like feet this would cause the animal to move on land in a rather awkward, jumpy manner (reminiscent of the animal seen on land near Sraheen's lough, Achill island in the late 1960's!).

Secondly *Potamotherium*'s deep chest, long neck, sleek profiled head, visible external ears and long muzzle are uncannily similar to both the animal portrayed on Grace Connolly's gravestone of 1722 and eyewitness descriptions of the Sraheen's Lough animal (John Cooney and Michael Nulty described a greyhound or sheep-like head on a swan-like neck!)

The only major problem with reconciling the Dobhar-Chu with *Potamotherium* (and vice-versa) is size.

From eyewitness accounts of the Sraheen's lough animal it was immense 8 to 12ft long (2.4 to 4m) "bigger than a horse" was how Gay Dever put it. However, *Potamotherium* only attained the mundane length of approximately 5ft *(1.5m),* which would place it well within the size range of modern day species (the only exception being the saro or giant river otter of South America) -which can attain the incredible length of 7 to 8ft (2.13 - 2.4m).

If, however, the Dobhar-Chu was/is a descendant of *Potamotherium* (and there is no suggestion that *Potamotherium* itself would have survived unchanged) then the *38-25* million years or so of evolution would possibly have changed it in some distinguishable way. One such way could be the animal exhibiting gigantism and therefore significantly greater in size than it attained as evidenced by its remains in the fossil record.

If one day a giant marme otter larger than the living saro in length were discovered in the fossil record in this same region then there would be a precedent for the existence of such a beast and consequently its inclusion in the folklore and legends of the two countries.

Literary sources cannot be ignored (the Dobhar-Chu laying dormant in early references).

Indeed if anybody were to question the position of the Dobhar-Chu in Irish Mythology before its apparent and indeed infamous debut at Glenade Lough in 1722 then it is significant to mention that a man travelling around the west of Ireland by the name of Roderick O'Flaherty, whilst collecting stories for his book - *A Description of West or H-Lar Connaught* in 1684 tells of the Irish Crocodile.

He was writing about the interesting features of Lough Mask (a large Lake which is nearly connected to Southern Ireland's largest - Lough Corrib in Co. Galway).
He states:

"There is one rarity more, which we may call the Irish Crocodile, whereof one as yet living about ten years ago (in 1674) had sad experience. The man was passing the shore just by the waterside and spied far off the head of a beast swimming, which he took to be an otter and took no more notice of it, but the beast it seems lifted up its head to discern whereabouts the man was then diving swam under the water till he struck ground, whereupon he ran out of the water suddenly and took the man by the elbow whereby the man stooped down and the beast fastened his teeth in his pate (head) and dragged him into the water, where the man took hold of a stone by chance in his way, and calling to mind the knife he had in his jacket, took it out and gave a thrust of it to the beast which thereupon got away from him into the lake.

The water about him was all bloody, whether from the beasts blood or his own or from both he knows not. It was the pitch (colour) of an ordinary greyhound, of a black slimy skin, without hair as he imagines.

Old men acquainted with the lake do tell there is such a beast in it, and that a stout fellow with a wolf dog (Irish wolfhound) along with him met the like there once, which after a long struggling went away in spite of the man and his dog and was a long time after found rotten in a rocky cave of the lake when the waters decreased (level of lake fell).

The like (kind) they say is seen in other lakes in Ireland, they call it Dovarchu i.e. Water dog or anchu which is the same.

The above description of this creature is of an animal unlike the common Irish subspecies of Eurasian otter, for although it initially looked like one, it evidently was not one as it swam closer to its observer and unfortunate for the person involved it deemed them a very suitable item for dinner!

It is also very unfortunate for cryptozoologists and indeed anybody intrigued (such as myself) by the 'legend' of the Dobhar-Chu that there was allegedly a body of such a beast in a cave of Lough Mask related to in the previous account. Although this body, was available for scrutiny it was obviously in some state of decomposition (due to the adjective 'rotten' being used) and therefore would possibly not have been foremost in the minds of the local populace to keep it or accurately describe it for future posterity.

The main reason for this was that they were obviously fearful of it and would therefore certainly have not had the caring and mindful nature to have wanted to examine its carcass in order that future generations could discern its identity and reap any benefits. It may be very optimistic also to assume that its bones have survived to the present day. The exact location of the cave would not only have to be discovered, but would have to be above the water line of the lough for any impending search to be successful.

However if they were ever found then hard physical evidence of the Dobhar-Chu and its identity would be obtained. It is also interesting to note that in O'Flaherty's account there is more than one reference to an attack by the beast (the Dobhar-Chu), which to me is indicative of more than one animal and representative of its population status, even back then! When researching an article such as this, which contains both mythological and cryptozoological aspects, it is crucial to include any literary sources, which may assist in solving the mystery/ mysteries concerned. In Peter Costello's excellent book - *In search of Lake Monsters* published in 1974 the author includes two instances in his chapter dealing with the Irish phenomenon entitled Pooka and Piast, where known and unknown animals prove to have further reaching consequences than initially thought.

The first of these concerns indirectly the three priests' (Fathers Matthew Burke, Daniel Murray and Richard Quigley) sighting of an unusual animal they had seen whilst out on Lough Ree (one of Ireland's largest lakes and the centre of the three lakes of the River Shannon - Ireland's longest river) on May 18, 1960. Before the publication of his book, Peter Costello was contacted by a woman called Anne Kinsella from Gory, in Co Wexford. This woman must have communicated with the author more than once as he refers to her as a "correspondent of mine". He presumably would have accepted her information as reliable.

Very tantalising is the occasion when she writes to Peter Costello on July 5, 1963, that her brother who lived close to Lough Ree was quite unimpressed at all of the press coverage and attention the three priests got concerning their sighting. It transpired that her brother (his name is unknown) had shot otters sometimes 7ft in length at a lake situated close to his house (he lived near Lough Ree). Anne Kinsella's brother also noted that otters can move without disturbing a lakes surface and that the priests' description and subsequent drawing of their creature proved that otters were the identity of the mysterious monster.

Peter Costello himself comments rather briefly on the enormous size of the otters killed by Anne Kinsella's brother. He remarks that this could be an explanation for the Master-Otter of legend. Indeed the longest authenticated length attained by an otter in the UK or Ireland has been 6ft (1.83m) as referenced in Ronald Binns' book *The Loch Ness Mystery Solved.*

Once again, we are presented with a tantalising piece of vital evidence for the physical reality of the Dobhar-Chu however in best cryptozoological tradition no more information or actual remains seem to have been forthcoming concerning this incident. Taking into consideration the location where this man lived close to the lough where the priests sighted their unidentifiable animal then it is indeed very unfortunate that any further details are lacking.

Also to be found within the pages of *"In Search of Lake Monsters"* by Costello is an intriguing incident and account occurring in the same geographical area as that of Glenade Lough where Grace Connolly met her grisly death. It was told by the noted Irish story collector Sheridan Le Fanu. He mentions an attack on a young boy at a lake situated between Loughs Bran and Glencar by a feared beast known as the "dreadful Wurrum". This creature lived in this lake (frustratingly once again its name does not seem to have been recorded) and beared some physical resemblance to a donkey.

Glencar Lake is located in County Sligo bordering County Leitrim. It is only one mile distant from Glenade Lough where the legendary Dobhar-Chu killed Grace Connolly. In view of the two lakes' close proximity to one another could Sheridan Le Fanu have documented a possible attack by a Dobhar-Chu?

Although this story is inconclusive and lacking in much detail, it is nonetheless, thought provoking. Its only crucial fault would be reconciling the beast's label of 'Wurrum' with that of the well-known name of water dog or Dobhar-Chu. The usage of Dobhar-Chu for the Irish subspecies of otter is nowadays overshadowed by the term 'Mada Uisage'. Both terms literally translate as 'water-hound' or 'hound of the flowing waters'. Also mentioned by Le Fanu, is the usage of 'anchu' for the creature responsible for the incident and attacks at Lough Mask involving the 'Irish Crocodile'. Therefore the noun 'Wurrum' does not seem to be synonymous with the term Dobhar-Chu and animal behind the name.

Indeed, the supposedly fearsome (and not entirely) mythical beast which legend and folklore cites existed must have been relatively common during the 15th and 16th Centuries for the aforementioned accounts to have been noteworthy enough for them to be remembered - not only by the local populace - but also literally by Le Fanu and O'Flaherty.

Whilst researching the legend, I was very intrigued to discover a chapter on Achill Island in a book entitled - *"Ireland's Islands"* written by Peter Sommerville-Large - in fact the information about Achill is sourced from a book - *"Achill Island"* by Theresa Macdonald. In this chapter the author talks about the early history of the island and of the early trips made by visitors in the 1800s. He mentions the region's superb natural beauty, impressive and unique flora and fauna, and also of other attractions for the visitor. Such attractions included the famous sealcaves. One such cave was known as the 'Priests' cave"(which was probably aptly named due to the persecution of Irish Catholics during times of reformation when priests were forced to say Mass in unusual/adverse circumstances).

According to an old man by the name of Harris Stone, who lived in Dooaghart (around 1906) he told of a rare species of sea otter, which lived in a stagnant pool in the cave known as the Priests' Cave which had ferns growing around its entrance. The cave was to be found on the north slope of Slievemore, but sadly the cave entrance is no longer accessible to the sea due to erosion and is very dangerous because of this.

Furthermore, this sea otter could be distinguished by its unusual colour - black or almost black with a visible white patch on its throat. The above information is intriguing to say the least - could the early people of the island have known of a very different and possibly unique species of otter, and have deliberately incorporated it into their folklore and oral traditions? Also noteworthy is the very mention by the author of a 'sea otter' - there are no sea otters (*Enhydra lutris*) residing in Europe, let alone Ireland.

This 'sea otter', as mentioned by the elderly man Harris Stone, must have at one time been indigenous to the island and surrounding regions for it to be included in the oral tradition and folklore on Achill. The only other possible explanation for a 'sea otter' to be part of the Islands folklore would be sightings and known behaviour of the common and endearing Irish otter. Both the Eurasian otter and the Irish subspecies (which is slightly darker) do indeed inhabit coastal regions of both Scotland and Ireland.

Local folklore and beliefs might have incorporated this known behaviour (intentionally or unintentionally) into the legend of this 'sea otter' residing in particular the area where the seal caves are located. Even the black colouration of Harris Stone's sea otter can be attributed to one of the physiological characteristics which early zoologists endeavoured to procure separate full species status for the Irish form of Eurasian otter. Some specimens captured in the last 150 years were indeed totally black, leading erroneously to early zoologists proposing the Latin name *Lutra lutra roenisis* for its full title as a separate species.

Nowadays, the Irish otter is recognised as a distinct subspecies due to the darker colour of its coat and the near absence of white on its throat. This colouration of its pelt is distinct from the British subspecies. The rash conclusion to name the Irish form as *Lutra lutra roenisis* was after several black specimens were captured at Roe Mills in Co Derry. It is very crucial to mention that if the information concerning the 'seal-caves' and the rare 'sea-otter' was gathered by the author Theresa Macdonald, independently from any articles concerning the Sraheens Lough monster, then this is very significant evidence with regard to the animal seen in, and around, the lough (especially the flurry of sightings in the late 1960s). It also indicates that the strange animal was/is indigenous to the island and environs and that its existence was well recognised by the islanders before the incidents at Sraheens Lough in the late 1960s. Also to call it a 'sea-otter' must imply that it was an otter and not something else, as the local populace would be well acquainted with their native fauna, unless of course it very closely resembled one perhaps very superficially.

Is the Dobhar-Chu resident in a pub in Co Mayo?

Whilst travelling to Connemara during April 1999, to do some research on the horse-eels seen in the region's lakes - or loughs as they are more commonly known in Ireland - I stopped with my companions - my wife (Bridget Cunningham), grandmother (Eilish Donnan) and aunt (Louise Donnan) at a small village called Crossmolina which is in Co. Mayo. This village is approximately five miles from Ballina, and is located on the shores of a large lake - Lough

Conn.

My grandmother and wife told me that there was a large stuffed otter in a pub they had visited (in order to use washroom facilities). This pub is called Hynes pub and was decorated with old photographs depicting major events of local history dating to the turn of the 20th Century. It also possessed many stuffed animals including amongst them the otter that my grandmother had noticed - being aware of my interest in Irish Lake monsters and also of the Dobhar-Chu legend.

The otter was located in one corner of the pub and was positioned above a television. It was very dark in colouration - almost appearing black – and was approximately four and a half feet (135 cm) to five feet (1.5m) in length. What grabbed my attention even more so than its size was its unusual shape. Its neck seemed unusually long and sleek, its hind legs were slightly elongated, and - most eye-catching of all - it possessed a long thick bushy tail!

There was also a distinct lack of white visible on its throat patch. The overall appearance of this otter was unlike that of the Irish otter, and differed markedly.

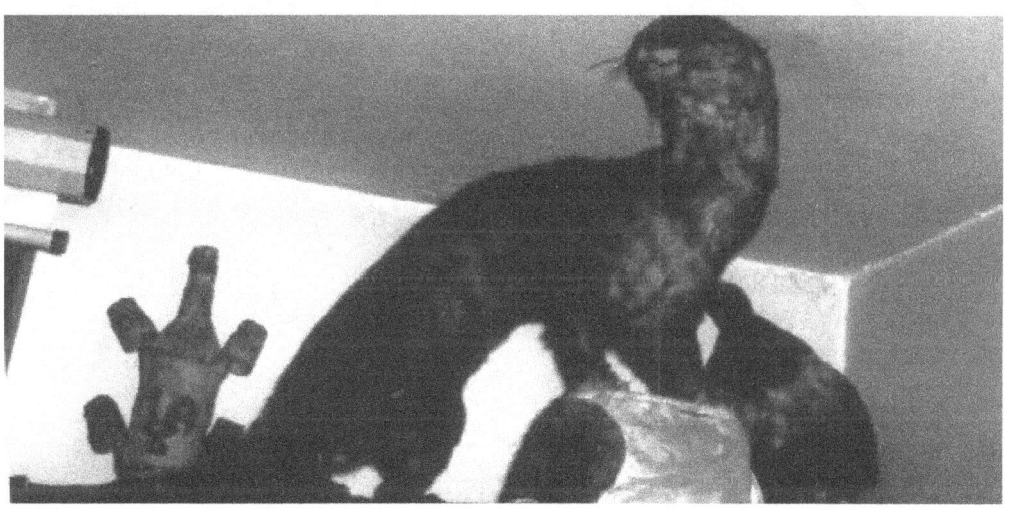

Unusual Otter specimen in Hynes Pub, Crossmolina Co Mayo: Could this be the real animal behind the legend? Note the lengthy neck, supple and lengthy body and tail. This gives the animal a very sleek profile. Approximate estimated length four and a half to five feet (135-150cm). Origins unknown. Note the distinct lack of white on throat. Do even larger specimens (perhaps 7ft or longer) exist somewhere in Ireland in stuffed or living form?

Images of a conventional Eurasian otter, together with another image of the mounted otter in the pub, (below) for comparison

Above is a photograph of a stuffed otter owned by a friend of mine - Neil McCooey. After telling Neil about my interest in the Dobhar-Chu and mythology associated with it, Neil was very keen to tell me of the large specimen he had shot at a lake near Cullyhanna, South Armagh sometime in the 1970s.

The otter is approximately between four and four and a half feet (120-135cm) in length and Neil informed me that it was female (a bitch otter). The original colour of its coat has faded due to the harmful ultraviolet rays of the sun over the past three decades (its original hue was apparently a rich chocolate colour).

Neil told me that his otter was very powerfully built and heavy (in fact his spaniel retriever dog could barely carry it back to Neil in its mouth) and this in conjunction with its overall morphology provides a direct comparison with the Hynes pub specimen as previously discussed.

For example the specimen that Neil owns from South Armagh has the normal hind leg shape and proportion for the Irish subspecies of European otter (*Lutra lutra roenisis*), whereas the Hynes pub specimen possesses uncharacteristically long hind limbs with a more pronounced knee joint (as can be discerned from the second photograph of it). The South Armagh otter also has the shorter tail unlike the Hynes pub animal which has a far thicker, bushier and presumably more powerful tail. Apart from the obvious difference in colour (a point which should not be dwelt upon due to the destructive ultraviolet rays of the sun) the other main physiological difference would be the more robust shape of the South Armagh specimen as opposed to the stream lined shape of the Hynes pub animal.

The Hynes pub animal also has a more lengthy neck. All of this brings me to the conclusion that the specimen in Hynes pub, Crossmolina, Co. Mayo bears not only an uncanny resemblance to the animal portrayed on Grace Connolly's gravestone but also that it shares some features with the prehistoric otter – *Potamotherium*. And, as already mentioned, this fossil otter was more suited to life in water than on land - this would explain why the Dobhar-Chu/ Sraheen's Lough animal seemed awkward whilst moving about terrestrially.

In fact the Hynes pub animal's shape would enable it to possess better hydrodynamic performance m it's watery medium and conversely why it seems clumsy on land.

From the longer hind limbs and larger, more powerful, tail of the Hynes pub specimen it is tempting to suggest that it may be exhibiting certain morphological traits from its evolutionary past caused by a recessive gene occurring during intermittent years. This might account for the Dobhar-Chu being regarded in mythology as the "seventh otter of a seventh otter". Thereby reflecting the irregular appearance and uniqueness of a very special individual/s in certain years.

This situation is not unique and has a parallel with another once mysterious and elusive mammalian carnivore - the king cheetah of Africa. Giant cheetahs with stripes instead of the spots of the regular African cheetah were reported throughout the southern part of the continent over the decades. However it wasn't until Paul and Lena Bottriel searched for concrete evidence of its existence by way of a corporate sponsored expedition in 1975 that they were successful.

As predicted by the "father of cryptozoology" - Bernard Heuvelmans - the king cheetah proved to be nothing more than a very unusual variation of the regular cheetah caused by a genetic mechanism resulting in the animals exhibiting their striking markings. Indeed it may be that the king cheetahs are evolving to cope with the environment in which they live (they prefer forested areas whereas the normal African cheetahs prefer open plain and savannah - the king cheetahs stripes provide it with better camouflage than the spots of the normal cheetah).

So in fact the Dobhar-Chu could indeed have been a variation (albeit a very unusual form with striking morphological and behavioural differences - such as gigantism and a pronounced carnivorous tendency leading to far different dietary needs than the normal species).

The common Irish subspecies of otter seems to have been quite capable of producing a radical and highly unusual variation i.e. the Dobhar-Chu and it may not even have been a new subspecies or species in its own right.

Can hybrids offer up a solution?

Whilst researching the Dobhar-Chu legend I was introduced by a friend of mine - Neil McCooey - who knew of my interest in the legend, to a man called Eddie Lennon.

Eddie, who is originally from Poyntzpass, Co Down and now lives in Dublin, is a retired dog breeder once specialising in the Irish water spaniel (which is of ancient origin and is the tallest of the spaniels). Eddie has future plans to include and incorporate the legend, and animal responsible for it, in a book he is planning on the Irish water spaniel.

He informed me, in November 2000, that the local population of County Roscommon believed that the origins of the Dobhar-Chu were due to chance and indeed occasional orchestrated encounters between a female water spaniel and a male or dog otter. Local tradition states that the bitch spaniel was transported to an island known as Otter Island in Lough Ree (a very large lake in Southern Ireland and scene of the three priests' sighting in 1960) and subsequently abandoned.

Nature then played its part with the successful liaison between the two animals said to produce offspring, which was reputed to be that of the master otter/Dobhar-Chu of Irish legend. If this intriguing event did somehow occur and, more importantly, was biologically feasible, then this hybridisation would indeed account for both the dog-like and otter-like physiological features of the Dobhar-Chu.

However, the resulting hybrid would be phenomenal in itself as the successful mating between two species of animal requires those species to be from similar animal families. And as otters are from the Mustelidae and dogs from the Canidae groups of mammals, then the zoological consensus would be that such an encounter would not produce viable offspring, however stranger things have happened in the world of nature and zoology.

If it was indeed possible and feasible then there are certain physical characteristics possessed by the Irish water spaniel that would account for the resulting morph/hybrid displaying features exhibited by the Dobhar-Chu and more specifically the image of it portrayed on the tombstone of Grace Connolly.

Its pointed muzzle, powerful arching neck, barrel shaped ribcage (which accentuates the spaniel's rolling gait - reminiscent of the movement of the animal seen at Sraheens Lough), and its straight powerful hind and forelegs, are all evident on the depicted creature featuring as a gruesome epitaph on Grace Connolly's final resting place.

However there is one additional point regarding hybrids - the animals nearly always tend to be infertile and thus incapable of producing a viable breeding population enabling the animal to survive!!

Nothing ever stays the same!

Why there are no apparent post 1968 sightings of animals similar to the Sraheen's Lough monster and ultimately of the Dobhar-Chu itself?

I believe that the most tragic strand of the Dobhar-Chu legend is the tragic fact that no creatures resembling it, such as the Sraheens Lough monster, have been reported from Ireland since the events on Achill Island in 1968. The reasons for this inevitability may be as follows.

Initially the animal (or animals) could have been endemic to the island (as the story told by Harris Stone of the rare 'sea-otter' indicates) or attracted to it by the presence of fishing activities and therefore the abundance of an easy, nutritious food supply i.e. salmon and sea-trout. In fact the Sraheen's Lough animal may have been a transient visitor to Irish shores and that the entire ecosystem, which existed on Achill Island before 1970, was the reason why it stayed or possibly visited annually.

Sadly, however this once delicately balanced ecosystem has been tainted with the ever-increasing demand for tourists seeking isolation, sheer natural beauty, water sports and recreational pastimes such as walking and fishing. This conversely has led to increases in more population, greater traffic flow and chemical pollution emitted by this traffic, and a subsequent reduction in areas of unspoilt wilderness as holiday homes and accommodation become a priority in order to assist with the island's growing economic prosperity. Noise pollution is the most influential and destructive form of pollution regarding the absence of the Sraheen Lough monster and animal responsible.

However, pollution caused by farming, for example infiltration of fertilizers and nutrients into the food chain and hydrographic system (lakes, streams and rivers) would be virtually non-existent on Achill due to the poor land and consequent lack of intensive farming practice.

In fact, the greatest attractant for any carnivorous animals to the island i.e. the plethora of fish and traditional migrational routes of salmon, sea trout are now in peril themselves. Fish farming of salmon in order to cater for the increasing demand by consumers can create major problems, which alter the biological food chain. Sea-lice and disease from cultured salmonoid species due to build up of faecal material can eventually filter into an already delicately balanced ecosystem. Thus natural stocks become decimated by disease and parasites (such as the sea-louse).

Therefore any animal returning to once traditional safe sanctuaries would discover a very different environment particularly with regard to the availability of prey species and also any truly isolated areas, suitable for breeding or giving birth remaining.

Another possible reason why sightings of the Sraheens Lough animal post 1968 have been non-existent could be that the animals themselves are scarce. Predators at the top of the food chain due to their very nature tend to be scarcer than their prey so in fact the animal(s) responsible for the Sraheen's lough sightings may have suffered a severe decline in numbers or died out completely or beyond the point of irreversible extinction inconceivable as that would be.

If, however, the creatures accounting for the sightings were merely visitors to Achill Island and not endemic, then they may have sought sanctuary elsewhere - possibly on another remote island habitat or coastal shoreline of a country further north towards the Arctic Circle.

It may even have sought sanctuary further west towards the lands of Iceland and Greenland where a similar animal may be incorporated into their mythology and folklore. All of this is, of course, mere speculation, but if a specimen similar in body shape to the Hynes pub animal, only darker in colouration and more significantly far greater in size (in particular lengths perhaps 7ft or longer) was discovered then this may conclusively prove the physical reality of the Dobhar-Chu/Master otter of Irish legend.

Zoologically and scientifically the discovery of such an animal would prove fascinating as it would demonstrate how a relatively small island country (small in global terms) such as Ireland could harbour such a previously undiscovered giant species - if it were indigenous to Ireland and not a migratory species from elsewhere.

The specimen would, of course, have to be offered up for formal scientific scrutiny if one is ever to 'fall into' the right or possibly wrong hands, as exploitation can become an inevitable reality of an endangered or new animal as it becomes available to science.

Bellacragher Bay, Near Mallaranny on mainland. This bay is not too distant from Achill Sound and Achill Island itself. The practise of salmon fishing is evidenced by the presence of nets and equipment. This was undoubtedly not the case during 1968 and before.

THE DOBHAR-CHÚ (MASTER-OTTER)

- endemic at one time to the west of Ireland (and midlands)
- extinct circa 1970.
- stuffed specimen may exist in Hynes pub Crossmolina, Co. Mayo
- length 6-8ft (possibly as long as 12ft!)
- preference for coastal areas
- strictly carnivorous
- stronghold near Achill Island, Co Mayo and counties Sligo and Leitrim.

- THE ABOVE IS MY OWN ARTISTIC RENDITION OF THE DOBHAR-CHÚ BASED UPON THE ASSUMPTION THAT IT WAS /IS A GIANT SPECIMEN /SPECIES

G. Cunningham '00

Glenade Louch, Co. Leitrim: This picturesque and idyllic lake was the scene for possibly Ireland`s most mysterious and only murderous attack by a lake monster – the otter like Dobhar-Chu

Gone but not forgotten?

Conclusions and thoughts on the possible outcomes of the mystery.

When we are searching for the truth behind a mystery, which has cryptozoological implications, then we must consider the notion that local folklore has created the reality for such legendary animals. This 'folklore muddle' whereby the known characteristics of one or more animals have been lumped together to form a composite animal (one which is not real in the strict sense of the word) may have happened in order to try and explain a tragic event i.e. the attack and subsequent death of Grace Connolly at Glenade Lough by an animal very mundane - perhaps a feral dog.

Such a composite animal would then be subjected to the tall tales traditionally reserved for the male sector of society. Over successive generations and slightly differing accounts each time an animal would be created radically different from that which was responsible for the story in the first instance. This situation is akin to the phenomenon known as 'Chinese whispers' when an actual event is related orally within society after each successive recounting of the event there is a slightly differing account of it.

The effect of this is that the entire story 'snowballs' and over many generations the final version of the tale is quite unlike the original and factual one! Even with advancements in science, technology, industry and society itself, Ireland has not suffered the population explosions, which have happened in other European countries and indeed around the world. Consequently, there are still many regions in Ireland, which are infrequently visited and free from man's exploitation.

Perhaps in areas of Counties Mayo, Sligo, Galway and Donegal specimens of giant otters still exist unnoticed by man due to their nocturnal habits and wariness from man's attentions. Examples of these unspoilt wildernesses include Doo Lough situated between the Sheefry Hills and Mweelrea mountains in Co. Mayo.

Incidentally, this lake is located in the same geographical area and only 20 miles distant from the village of Crossmolina where the unusual otter specimen resides in Hynes Pub. Doo Lough is a relatively small lake some *2.5* miles in length and *0.5* miles in width and is visited by farmers, fishermen and walkers, albeit infrequently. Another example of an area suitable for a giant species of otter to exist unthreatened by man is Lough Carrowinore near the peninsula known as The Mullet in north western Mayo and also Lough Muck in Co. Donegal.

This latter lake was the setting for a water monster with 'big eyes' seen by a woman picking bog-bean at its shores in 1885. I have visited this small lake and it is well hidden from the main road the R250 from Fintown to Glenties. The surrounding area is hilly and forested and is relatively unpopulated, so maybe specimens of the Dobhar-Chu still exist here or frequent the place from time to time.

One also has to remember that due to the animal's quadrapedal disposition it could be more than capable of migrating from lake to lake when the need arises e.g. because of disturbance by man or shortage of available food. This ability is one of the main reasons why sceptics of the Irish Lake monster falter. The sceptics state that no animal could reside permanently in a tiny lake such as Sraheen's Lough and that any available food supply i.e. endemic species of fish would be depleted in no time. However, the animal's adeptness at negotiating and traversing varying types of terrain would ensure that it does not become isolated and suffers as a result. Also noteworthy is the topography of Achill Island and its environs, which is very suitable for any animal(s) wishing to remain undetected by man.

A multitude of bays, smaller off-shore islands coupled with the mountainous terrain and abundance of small lakes all help create an area which would enable an already endangered species safe sanctuary from man. To return to the physical features of the animal or animals responsible for the Dobhar-Chu it is significant to reiterate the creature's large body size.

Heat-loss by an ocean dwelling mammal such as the Pacific sea otter is crucial if that mammal lives in a very cold environment such as the Pacific Ocean. One evolutionary adaptation, which this mammal would possess in order to combat this, is body size (in their own tasonomic group - the Mustelids). Therefore the trait of gigantism which is very evident in the Brazilian giant river otter, lends much credibility to the notion of a giant, as yet undiscovered species or subspecies of otter surviving (historically at least) in western counties of Ireland.

The very definition of cryptozoology - the scientific study of undiscovered animals unexpected in shape or form but also in time or place is very appropriate for the entire Dobhar-Chu mystery and creature(s) involved. Dr Karl Shuker's article for *Strange Magazine #16* deals with the mystery very accurately and cohesively mentioning the historical perspective and accounts within antiquarian articles in Ireland. It also links the artist's reconstruction of the Sraheen's Lough monster (as appeared in the part work - *Creatures from Elsewhere* published in the 1980s and edited by Peter Brookesmith) with the animal depicted on Grace Connolly's tombstone. He also devotes a chapter to the legend in his recent book entitled - *Mysteries of Planet Earth*, which includes an artist's reconstruction.

This portrait evoked an important image of the animal as its coat is ermine like and quite convincingly enables it to be placed in the Mustelid group of mammals (if a specimen of the Dobhar-Chu was made available). It is very relevant to note that even though my theories proposed in this article are speculative and some may be very difficult to prove, I feel that they are nonetheless the best available when we consider the available anecdotal and literal evidence.

The main aim of this article is ultimately to generate discussion and further research into the mystery and from a wider context the Irish lake monster phenomenon in general. Only by doing so is there a glimmer of hope in attempting and succeeding in solving the identities of these elusive, unique and - sadly - rare or possibly now extinct inhabitants of Ireland's loughs and wilderness areas.

In conclusion, the mysterious legend itself and true identity of the creature(s) may never be satisfactorily solved. The possibility that someone in Ireland possesses a stuffed specimen of the Master Otter is believable and cannot be ruled out when we consider that the practice of stuffing animals that were shot/killed was commonplace in bygone eras. Perhaps some old man does have a strikingly different (and possibly unique) otter, immense in both size and weight and reaching a length of 7ft (2.1 3m) or more stuffed by a taxidermist and displayed in the house his family have occupied for generations.

He may indeed have recounted tales associated with it many times before. From a pessimistic point of view, we may never find such a specimen, as it may already be moth-eaten, its true colour faded by the harmful rays of the sun, and it may then have been discarded many decades ago and long since forgotten about! As discussed earlier, I firmly believe that a quite plausible explanation for the taxonomic identity of the creature will be discovered whenever a giant marine otter excelling even the living saro in size is found one day in the fossil record.

If the true identity of the Dobhar-Chu is indeed a member of a prehistoric mammal lineage then we will have unmasked one of my country's most enigmatic and yet apparently lethal lake monsters. A very compelling and fascinating footnote to the fabric of the entire legend occurred when I spoke about the subject last November to now retired Irish water spaniel breeder Eddie Lennon.

Eddie informed me that whilst conducting his own research at Grace Connolly's tombstone at Conwall cemetery in the town land of Drummans Co.Leitrim, a woman approached him and began to talk about the legend. The local woman then proceeded to tell him that the people of Conwall and neighbouring town lands intended to stage a festival sometime in the near future to commemorate the Dobhar-Chu legend. Given that the county is perhaps the sole region in Ireland, which still attaches a strong intrinsic belief in the legend, then her intended aspiration will undoubtedly assist in keeping the story and folklore of Grace and her unfortunate (and presumably ghastly) fate at the jaws of one of Ireland's most elusive mystery denizens of its loughs alive for many generations to come!

SELECTED BIBLIOGRAPHY AND FURTHER READING

1. *The Velvet Claw - A natural history of the Carnivores* By: David Macdonald (BBC Books 1992)
2. *Achill Island*- By: Theresa Macdonald
3. *Strange* Magazine, Issue no.16, pg32, 33 & pg 49-   Menagerie of Mystery regular feature (then!)By: Dr Karl Shuker. Sadly this excellent objective magazine is no longer in print!
4. *Mysteries Of Planet Earth - An encyclopaedia of the Inexplicable*- By: Dr Karl Shuker, pg 172 & pg 173 (Carlton Books 1999)
5. *Ireland - A Natural History- By: David Cabot* (Harper Collins 1999)
6. *The Age of Mammals*- By: Bjorn Kurten

ACKNOWLEDGEMENTS

This article would not have been possible if it were not for many people. I would especially like to thank my father and mother - Jim and Deirdre Cunningham for constant encouragement, support and instilling in myself that "accomplishment is best achieved through diligence" and for their unconditional love.

In particular this work would not have been possible if it were not for the very kind assistance from fellow lake monster researchers Peter Costello and Nick Sucik - who I am indebted to for sending me difficult to find early references relating to the Dobhar-Chu and use of copyright material from major works/books penned by them.

Heartfelt thanks also go to Jonathan Downes and Dr Karl Shuker for continued guidance and patience when bombarded by my many questions and conversational topics. Ronan Coghlan and Excalibur books, Glynn Watson of Glynn's books and Stephen Shipp from Midnight books have been instrumental in enabling me to continue researching the Irish Lake Monster phenomenon and cryptozoology in general - very sincere gratitude to all of them.

My work colleague and friend Michelle Murphy must receive special mention and praise as this article was drafted and printed several times before I think I have got it right! My grandmother Eilish Donnan deserves special mention not least of all because she is always there for all of us and was a valued proof-reader.

Last, but not least, immeasurable gratitude must be reserved for my loving wife Bridget. She has, I feel, over the years become very accustomed to unscheduled deviations from the regular holiday itinerary when I have been near (or sornetimes not!) an area of lake monster activity and particularly so when there has not been a variety of shops for her to explore and indulge herself in.

All possible effort has been made to contact all relevant persons concerned for permission to use copyright material and illustrations; unfortunately this has not been possible in all instances Any future editions of this work will amend this.

The Complete Mystery Reptiles

EDITOR'S NOTE: The following seven articles were written for the now defunct magazine *Pet Reptile* during the first months of 2001. Because the magazine was not widely available I received a large number of requests for photocopies of the articles and so it seemed appropriate to reprint them here in full

MYSTERY REPTILES OF THE WORLD: PART ONE
MONSTERS OF THE MILLENIUM

by Jonathan Downes

To some people it may seem inconceivable that at the beginning of the third millennium there are still zoological mysteries to solve. But it's true! New species of animals are being discovered each year, and although it is true that many of them are relatively small creatures, some significant zoological discoveries are still being made each year. Within the last decade, for example, over a dozen new species of sizeable land animal have been discovered in the war ravaged forests of Indo-China, and as we enter the third millennium it seems certain that many more exciting discoveries await us.

I am the Director of the Centre for Fortean Zoology (www.eclipse.co.uk/cfz) a non-profit making organisation dedicated to searching the world for new species of animal, and I have also kept exotic reptiles, amphibians and invertebrates for over thirty years. In this new series for *Pet Reptile* my colleagues and I shall be examining a wide range of mystery reptiles and trying to find out the truth behind the myth. Many of the most tantalising animal mysteries that confront us at the beginning of the 21st Century appear to involve reptiles. I am sure you have all heard about bigfoot, or the Loch Ness Monster, but what about the "Father of all the Turtles", or the Bujang Senang, the giant crocodile of Borneo? What about the mystery monitor lizard of Hong Kong, or the Tatzelwurm (a giant two legged skink said to live in the Alps)? These animals are very real and just as exciting to the Fortean Zoologist as any of the monsters better documented in the media.

There are even mystery reptiles in the United Kingdom! As I am sure every reader of this magazine will know, there are only three species of lizard known in these islands; the sand lizard, the common lizard and the slow-worm. However, for nearly two hundred years now there have been reports of bright green lizards, considerably larger than any of the native species which have been reported along the coast of East Devon and parts of Dorset. These animals are not just known from anecdotal reports but have been described in such august journals as the zoological society report from the Devonshire Association.

The scientific luminaries of the 19th Century were convinced that there was in fact a relict population of a fourth species of lizard living on the south coast of England, but after about 1920 everyone forgot about it until I rediscovered the old scientific papers a few years ago in a museum basement! After having investigated these reports for a number of years it seems (although I cannot prove it) that I am now pretty sure about what these creatures are.

There is a large and beautiful lacertid called the green lizard *(Lacerta viridis)* that is found across much of Europe and is even found in the Channel Islands. Until very recently there was a thriving trade between the Channel Islands and the Dorset port of Weymouth (where boatloads of tomatoes and other delicacies were often unloaded before being transported into the hinterland for sale). I am sure that it is not a coincidence that when one plots the sightings of these green lizards over the years on a distribution map that Weymouth proves to be the epicentre. This suggests that over a period of perhaps two hundred years *Lacerta viridis* has been an unwary stowaway to this country inside baskets of fruit and vegetables. Upon disembarking on the UK mainland they have spread into the surrounding areas where they have become established, and may even have bred.

Unfortunately, historical experiments in introducing this and other European species to these shores have always failed because of our cold winters. It is probable, I am afraid, that these intrepid lacertid travellers have succumbed to our inclement climate. Global warming, however, may bring us some benefits, because if the trade in fruit and veg from the Channel Islands continues, and the winters in the South of England continue to be as mild as they have been, then the beautiful green lizard may well eventually become an accepted member of the British herpetofauna.

MYSTERY REPTILES OF THE WORLD: PART TWO
The mystery monitor lizard of Hong Kong

by Jonathan Downes

Hong Kong is possibly the most enigmatic place on earth. A cultural, economic, political and ethnological cross-roads, it has been described as "a little piece of the Home Counties tacked on to the edge of Southern China", and it is also, undoubtedly the only part of mainland China to have escaped the ravages of fifty years of Communism. Despite its reputation as the world centre for *lassez faire* capitalism and 21st Century technology, the twin influences of British and earlier Chinese Imperialism were instrumental in producing the place we know today! Despite its high population density (six million or more people in an area somewhat smaller than the Isle of Wight), and its appalling problems with pollution and urbanisation, it has a rich and varied wildlife. Its position at the 'cross-over' between the tropical and Eurasian geographical areas has given it a unique fauna, which in many ways has a similar relationship with the zoology of the Pacific rim, as Hong Kong itself has with the socio-political and economic infrastructure of the same area.

The largest lizard known in southern China is the water monitor *(Varanus salvator)* which grows up to a length of between two and three metres. There are a few records from the paddy fields of rural Hong Kong but it is generally agreed that these are vagrant specimens that have wandered over from the mainland.

On the 21st January 1930 a lady walking along Lugard Road on Victoria Peak – the largest mountain on Hong Kong Island - was frightened when she saw what she thought was a "miniature crocodile". With the help of a passing policeman and some Chinese coolies, and a "Japanese gentleman who was passing" they cornered the creature. With great presence of mind the un-named Japanese Gentleman took off his coat and threw it over the animal. The lizard later allowed itself to be dumped in a sack and to be taken to a Police Station and ultimately to the Botanic Gardens where it 'was placed in a cage'.

It was examined by Dr Geoffrey Herklots, the most famous naturalist then living in Hong Kong. His description was:

Total length - 22 feet 10¼ inches, head: 6 inches, tail: 1 foot 6¼ inches.

Breadth - At neck 2¼ inches, middle of body 6 inches, in front of hind limbs 2½ inches, middle of tail 1 inch.

Depth - Base of tail 2 inches, groove along back and beginning of tail, ridge along rest of tail.

Colour - Above brown-grey, or deep olive, with yellow spots or hands, below a dirty yellow, neck no distinctive bands,

neck no distinctive bands,

Eyes - Open and close independently, lower lids move upwards. Iris a marbled pale Vandyke brown with a very narrow white or very faintly yellow circle immediately next to pupil.

Herklots noted that this was only one of several records of strange lizards seen both on Hong Kong Island and on the mainland at the time. As a child in Hong Kong I saw a similar creature on the Island during 1968. Various suggestions have been made for the identity of these creatures. It was initially identified as *Varanus bengalensis*, a species that isn't actually found in China. It was also tentatively identified as an African species – *Varanus albiguaris*. The surviving photographs, however, suggest that it was not either of these species. It is also certain that it is not the indigenous *Varanus salvator* so what was it?

Although these days exotic animals from all over the world are kept as pets, and escapees undoubtedly can and do become established in the wild, the international trade in exotic reptiles was almost non existent seventy years ago. Therefore, the suggestion that the lizard, which died soon after capture, was an escaped African *Varanus bengalensis* can, I think, be discounted.

Unfortunately the originals of the photographs were destroyed during the Japanese occupation of Hong Kong during WW2, as was the preserved body of the unfortunate reptile, and these rather substandard pictures are all that remain. Is there any monitor expert in the *Pet Reptile* readership who can help solve this seventy-year-old puzzle and set the mind of one curious cryptozoologist, at least, to rest?

MYSTERY REPTILES OF THE WORLD: PART THREE
THE FATHER OF ALL THE TURTLES

by Jonathan Downes

No-one would deny that there are still mysterious animals living beneath the waters of the world's oceans, which have not yet been described by science. In 1976 for example the world's third largest species of shark was discovered. Known as the megamouth shark because of its enormous mouth, this creature (which can reach a length of seventeen feet) seems to have a worldwide distribution.

Another enormous creature which has been reported from the entire world's oceans, but whose identity has not, as yet, been accepted by mainstream science is a gigantic chelonian, known to the natives of Sumatra as `The Father of All the Turtles`. Some religions even believe that the

world itself is carried on the back of a gigantic turtle (a concept taken by Science Fantasy author Terry Pratchett) and images of these turtles can be found in temples across the region where they are often used to hold up pillars.

Accounts of these remarkable animals first reached Europe in the early nineteenth century when Dutch settlers in what is now Indonesia reported native legends of enormous turtles. These stories eventually filtered back to their homeland in Western Europe. However, sightings of such beasts are not confined to the tropical waters of the East Indies.

In June 1956, seamen of the cargo steamer *Rhapsody,* reported that they had seen a huge turtle about 45 feet long with an all-white shell south of Nova Scotia. The Canadian coastguards warned local boats about this gigantic reptile with flippers 15 feet long and capable of raising its head 8 feet out of the water. Seventy-three years earlier, not far away on the Newfoundland Banks, a turtle 60 feet long and 40 feet wide had been reported.

What is really exciting is that these creatures have even been reported in British waters. In September 1959, a shark fisherman called Tex Geddes - who had once been an associate of renowned naturalist and author Gavin Maxwell - and James Gavin, a friend of his, who was on holiday, saw a giant turtle in the sea off the small island of Soay in the Inner Hebrides for an hour.

They had been watching marine creatures, including some killer-whales and a basking shark, when Gavin noticed a black shape on the water about two miles away in the direction of the Skye shore. Although this was where the killer-whales had last been seen, Geddes was convinced that this was something new. He later wrote:

"I am afraid we both stared in amazement as the object came towards us, for this beast steaming slowly in our direction was like some hellish monster of prehistoric times. The head was definitely reptilian, about two feet six high with large protruding eyes. There were no visible nasal organs, but a large red gash of a mouth which seemed to cut the head in half and which appeared to have distinct lips. There was at least two feet of clear water behind the neck, less than a foot of which we could see, and the creature's back which rose sharply to its highest point some three to four feet out of the water and fell away gradually towards the after end. I would say we saw 8 to 10 ft. of back on the water line."

Slowly the creature swam nearer, and the two men watched it until it was parallel to the dinghy only 20 yards away. It kept turning its head from side to side, as if looking all around it. Eventually it submerged and the two men headed gratefully for the shore.

One could hardly hope for two better witnesses. They were both experienced seafarers and fishermen who were familiar with the local wildlife. However these are only a few of many sightings of giant turtles seen in the world's oceans. What could they be?

Firstly, in the best traditions of fisherman's tales, I think that we can discount any suggestion that there really is a turtle with a shell measuring between forty and sixty feet in length. At sea, with no frames of reference, sizes are notoriously difficult to estimate and it seems certain that whatever it was that the crewmen of the *Rhapsody* reported seeing it was considerably smaller.

However there is a great deal of evidence for the existence of giant marine chelonians. In the cretaceous period (which ended 65 million years ago) there was a giant turtle called *Archelon.* It was found in the sea of Niobrara over what is now the state of Kansas in the USA. The carapace was twelve feet long and the skull was three feet long. Some zoologists have speculated that 'The Father of all the Turtles' is a surviving population of *Archelon.*

However it is not even necessary to hypothesise such a Jurassic Park type scenario in order to explain these magnificent creatures because there is already an animal very well known to science which could explain all these sightings.

The leathery turtle, or luth, is found in all the world's oceans and although it breeds in the tropics, particularly in Indonesia and coastal Central America it regularly visits temperate waters including Scandinavia, Nova Scotia and the north of Scotland. The largest specimen ever found was found in the 1980s near Harlech in Wales and was over nine feet in length. It is certainly not impossible that even larger specimens exist!

If one plots a graph showing the distribution curve of the size of animals one can extrapolate some interesting data. An average human male, for example is just under six foot tall. I am six foot six, and have known people who do not suffer from any genetic abnormality who are considerably taller than me! There are also perfectly normal blokes who are considerably shorter than the average. If one extrapolates this frequency curve for an animal like the leathery turtle then one discovers that if the average length is between seven and eight feet, an animal of nine feet is not uncommon, and that leviathans of twelve feet or even more in length are perfectly feasible.

The 'Father of all the Turtles' may not carry the globe on its back but certainly exists! The irony is that these magnificent animals are in danger because of thoughtless behaviour by human beings. Like their ancestor *Archelon,* these creatures' main diet are jellyfish, and it is a sad fact that many of these giant turtles die each year after eating plastic bags or toy balloons which have drifted out to sea after mistaking them for their favourite food.

Remember this, next time that you see a charity or promotional event where balloons are released willy-nilly into the air. Each one of these apparently harmless bags of rubber could be condemning 'The Father of all the Turtles' to an agonising and ignominious death!

MYSTERY REPTILES OF THE WORLD: PART FOUR
THE GAMBIAN SEA SERPENT

by Jonathan Downes

Over a decade ago, the blockbuster movie *Jurassic Park* captured the imagination of a generation. Although the scientific theorising was flawed, the idea that people might some day be able to visit a theme park full of prehistoric reptiles is an irresistible one. However, what few people realise is that some great creatures from the age of the dinosaurs may still live on the planet. One of the most intriguing mysteries of modern zoology is that of the Gambian sea serpent.

In June 1983, a British schoolboy called Owen Burnham - who at the time lived with his Missionary parents in the Senegal - was on holiday in The Gambia. He was a keen amateur naturalist, and had kept many of the smaller animals of West Africa as pets. They were holidaying at the Bungalow Beach hotel near Banjoul, and one morning Owen was strolling down the beach when he stumbled across a 15 foot long carcass that had been washed up by the tide. The creature was only recently dead and had not yet started to decompose, and resembled no known living animal. It had a long head with crocodile-like jaws, four turtle-like flippers and a slender tail with no fins. The skin was smooth and rubbery without scales. Unlike a dolphin or whale, the creature had no blowhole. It most resembled a pliosaur, a group of marine reptiles believed extinct for 65 million years.

Burnham, who had no camera, took detailed sketches and measurements that he later sent to *BBC Wildlife* magazine. Unfortunately, two locals came upon the cadaver and started to hack off its head to sell as a tourist souvenir. Knowing that it was potentially too valuable a specimen to lose, Burnham and his family helped employees from the hotel to bury the creature way above the tide level on the beach where the sand and hot sun would mummify the body. Burnham knows the exact spot where he buried the animal on the small beach.

He told Dr Karl Shuker:

"The creature was brown above and white below (to midway down the tail).

The jaws were long and thin with eighty teeth evenly distributed. They were similar in shape to a barracuda's but whiter and thicker (also very sharp). All the teeth were uniform. The animal's jaws were very tightly closed and it was a job to prise them apart. "

Representatives of the Centre for Fortean Zoology have been back to the Gambia four times. They have ascertained that the area of beach where Burnham buried the mysterious beastie so many years ago is happily untrammelled with buildings. They have photographed the area which is, ironically enough, only yards from the perimeter wall of the gardens of a popular tourist hotel. Feeling, quite understandably, that discretion was the better part of valour they have decided not to try and explain to machine gun toting Gambian policemen (who since the political unrest of modern West Africa are not exactly known for their tolerance towards interlopers) why they were digging an enormous hole in the beach without permission. Thus all excavations were carried out clandestinely and under cover of night, and were, not surprisingly, unsuccessful.

Now they are seeking funds to return with Burnham to the Gambia in order to dig up and examine the beast thus solving one of the most vexing riddles of modern zoology. We do not

even have to take the whole cadaver back to England; all we need are a few small bones for identification and some samples of flesh for DNA analysis.

Unlike many mysteries of Cryptozoology this one is potentially easy to solve one way or the other. We know the exact spot that the beast was buried. All we need is a couple of shovels and the mystery is solved. The animal involved here is dead, it cannot run away or hide so we do not have to waste time looking for it. Only a handful of people in the whole world know the whereabouts of the body; Burnham, his family and the CFZ, so the likelihood of it having been dug up is exceedingly remote.

Burnham says the local fishermen know of this animal. It occasionally turns up in their nets. They call it *Gnalo*. He also says a fisherman known to him has located the original skull taken from the animal in 1983. We could tour the coastal fish markets and interview fishermen who might have knowledge of the beast.

A scenario like this appears to have escaped from the pages of a 19th Century adventure novel by someone like Henry Rider-Haggard. Maybe it is time for a band of modern day Allan Quatermains to prove that whilst King Solomon`s Mines may have been a myth, the Dark Continent is even now a place of myth and mystery....

MYSTERY REPTILES OF THE WORLD: PART FIVE
THE GOLDEN FROGS OF BOVEY TRACEY

by Jonathan Downes

According to an old Devon folk story, once upon a time a poor woodsman lived with his family on the outskirts of the village of Bovey Tracey. Their child was suffering from an unspecified illness and was not likely to live much longer. One night, in the middle of a severe thunderstorm, there was a knock on the door and a mysterious lady entered demanding shelter and food.

Despite their many misfortunes, the woodsman and his family welcomed the mysterious lady, gave her milk and food (which they could ill afford) and a seat by the fire. She then blessed the ailing infant who was miraculously cured, and before vanishing (up a road called to this day Mary Street) she said that, so that her benefactors would know this was not a dream, not only would the child be forever cured, but that the next day the family would discover a new spring full of crystal clear water and bright golden frogs which were said to have populated the area for many years.

If so, what were they? The concept of brightly coloured amphibians inhabiting the English countryside is not as unusual as one might suppose. Several small pink frogs were found in Gloucestershire in the early 1990s and others have been recorded over the years from Sussex and The Cotswolds.

But golden frogs?

A stream running through the centre of Bovey Tracey

In February 1994, local and national newspapers were full of the story of Jaffa, a three-year-old frog discovered in a garden in Truro. Jaffa was, as his name implies, bright orange. The Westcountry TV News carried a story about him, which said that he, and a similarly coloured mate, had been released in a secret location. Mark Nicholson of the Cornwall Trust for Nature Conservation revealed that far from being an isolated occurrence these oddly coloured amphibians are popping up all over the place. Ranging in colour from bright orange, through yellow to pale cream, these creatures have been reported from all over the westcountry, and even from elsewhere in the UK - although they appear to be much rarer.

They turn up each year, and in the early spring of 2000 we found one hopping around in the mud by the dustbins outside our front door! It was a fully-grown female of a bright mustard colour and almost immediately laid copious amounts of spawn, which, unfortunately proved to be infertile. We kept her for several months until she escaped into the wilderness behind our conservatory where we keep some of our animal collection during the summer months. We hope that the coming years will supply us with more specimens that will help us determine what exactly these creatures are.

The Holy Well in Mary Street.

Although I am as aware as everyone about the damage that man is doing to the planet on which we live, there is a distressing tendency in these politically correct days to blame all anomalies of nature on the ubiquitous 'global warming' or some other plague of the modern age. Whilst not wishing for a moment to deny that these threats exist, and are very serious towards the future of all living creatures on the planet, it is, I think, counter productive to blame everything on man's stupidity. Various commentators, including Internet news groups, have suggested that these brightly coloured batrachians are the result of a damaged environment. Whilst not denying that the environment is certainly damaged, we have proved that these animals have been around for at least five hundred years and we are determined to find out what they actually are!

For the present however, it is fairly clear that the charming medieval legend of the Golden Frogs of Bovey Tracey might not be so far fetched after all.

MYSTERY REPTILES OF THE WORLD: PART SIX
THE MYSTERIOUS TEMPLE TURTLES

by Jonathan Downes

One of the central tenets of Buddhist and other oriental faiths is that all life is sacred, and over the years Buddhist and Taoist temples across the orient have become havens for wildlife of all kinds.

In Hong Kong, for example, where there is practically none of the original forest left, the *feng shui* woods which have been nurtured for generations around places of worship have become veritable oases for wildlife in a desert of urbanisation. However, in recent years naturalists have discovered that some of these temples have unwittingly become havens for completely unknown species of animals.

Recently, in Indochina, zoologists have been gathering reports of giant soft shelled turtles reported in sacred pools across the region. Perhaps the most exciting is one reported in Hoan Kiem Lake, in the Vietnamese capital Hanoi. This semi legendary creature, which has been photographed on a number of occasions, is reported to be six feet in length with a green shell and a pink belly.

"This turtle is a fascinating phenomenon, probably the biggest soft-shell in the world and certainly the most endangered," said Peter Pritchard, a renowned turtle biologist. "*People in Vietnam are treating it like the Loch Ness monster, but this is not a myth. People need to treat it like a biological thing -- an endangered species."*

The story goes that Le Loi, a warrior king, used a heaven-sent sword to hold off some Chinese invaders back in the mid-1400s. After the final battle, as Le Loi was boating in Hanoi, his sword leapt from its scabbard and into the mouth of a turtle. The turtle plunged underwater with the sword, never to be seen again, and the lake has been known as Ho Hoan Kiem ever since - the Lake of the Returned Sword.

However *Pet Reptile* can exclusively reveal that there is another potential new species of giant soft shelled turtle – in southern China. A recently discovered Chinese engraving from about 1840 shows what appears to be a Chinese religious procession venerating a giant chelonian. Although the creature is depicted as having a scuted shell, the shape of the face is so similar to known species of soft shelled turtles it seems likely that in view of the recent discoveries in Vietnam that more remarkable reptiles are lurking in the sacred pools of the orient.

MYSTERY REPTILES OF THE WORLD: PART SEVBN
THE SCARLET VIPER

by Jonathan Downes

I doubt whether there is a single reader of this magazine who doesn't know the answer to the next question. How many species of British snakes are there? *"C`mon Jon there are three!"* I can hear you shouting from your homes around the country, and you are right. Except that a hundred and fifty years ago, many people believed that there was a fourth species – *Vipera rubra* – The Scarlet Viper. Across areas of southern Dorset, especially those surrounding Lulworth Cove and Corfe Castle there were sightings of adders quite unlike their cousins recorded elsewhere in the country – they were bright red.

These animals were quite well known to scientists of the time and were mentioned in all the natural history listings. They were described as being slightly smaller than other adders, but were seen as being a distinct and very beautiful species. In the late 19th and early 20th Centuries the taxonomy of British wildlife underwent a massive shakedown period and quite a few previously distinct species lost their taxonomic status, and were relegated to being mere colour morphs or regional races of well known creatures. However it is the geographical location of the creature that has always puzzled me.

Relict populations of two mainland European species of reptile actually *do* live in that area of southern Dorset. The sand lizard and the smooth snake live here, and if a rare and hitherto unknown species of British viper were to be living anywhere in the UK it would probably be here. With the recent advances in DNA fingerprinting, it would seem to be a relatively easy task to find out whether the Victorian naturalists were right all, and the semi legendary scarlet viper of Lulworth Cove is indeed a distinct species after all. However, the problem is that we can't actually find a specimen....

Ironically, although the creature was a well-known one, and many specimens must have been taken for various collections, we haven't, to date at least, been able to find a specimen in any of the natural history collections. So friends. Do YOU fancy securing a little piece of zoological immortality for yourself? If you have the time or the energy pester your local museum or natural history society. Examine their records. Someone, somewhere must have a specimen of the elusive scarlet viper pickled in a bottle in their basement. All we need are a few tissue samples for DNA analysis, and we can lay this enduring and fascinating zoological mystery to rest once and for all.

Hybridisation in the family Felidae
by Chris Moiser

When looking at the reports of sightings of exotic or alien cats in the United Kingdom the question of whether they could be hybrids is frequently asked. In particular could there have been hybrids produced in the wild in the UK? And if so which species have hybridised and are they likely to continue to breed?

Unfortunately to answer these questions it is necessary to first consider some biological theory and a few definitions.

A species is normally defined as a group of individuals that have the potential to breed together and produce fertile offspring. Within the species we often have a number of sub-species that tend to be local variations on the species, these are usually separated from other sub-species within the species by geography, and so have adapted to the local conditions. Sub-species within the species tend to vary in colour, i.e. being lighter or darker than other sub-species, or in size; as a very loose generalisation sub-species on islands tend to be bigger than those on the mainland. Those that live in hotter climates tend to be lighter coloured. If brought back together the two sub-species (within the species) can breed and produce fertile offspring, these will usually appear intermediate between the two sub-species). If you follow the Darwinian theories of evolution the sub-species, if they remain separate, may continue to develop in their own particular ways until they eventually become separate species.

The word hybrid can have two different meanings; in the narrow sense it may be a cross between two individuals of the same species, but of different varieties, e.g. if a Siamese cat was crossed with a Persian cat, the offspring could, correctly be referred to as "Siamese/Persian hybrids". In the broader sense, the word hybrid is usually taken to refer to a cross between two closely related, but different species. It is, in the broader sense, that the term is used in relation to possible alien cat sightings.

In the wild hybrids between species almost never occur. This is for two reasons, firstly very similar species do not normally occur in the same geographic area, and secondly the behaviour of one species is sufficiently different from that of other species for the courtship behaviour not to be recognised. Even if a mating occurred during a period of receptivity fertilisation would almost certainly not occur because of a mismatch between the chromosomes of the potential mother and father.

This is a somewhat complicated mechanism to explain, but basically all mammals have their chromosomes in pairs, and one member of each pair comes from the mother, the other member from the father. When a female mammal forms eggs and the male mammal forms sperm the pairs of chromosomes line up along side each other and often exchange 'pieces' of chromosomes before forming the egg or sperm. The purpose of this is to mix the mothers and fathers genes within the chromosome that goes into the egg or sperm. This mixing is called "crossing over", simply because a fragment of one chromosome will appear to cross over to another. (Certain biologists have been known to refer to this as a "chromosomal leg-over"). The problem is that if the mother and father of the animal currently forming eggs or sperm are of different species the genes on their chromosomes (genes define individual characteristics in the animal) may be in different positions. Thus when "crossing over" occurs in the hybrid animal the egg or sperm it produces may be unviable simply because of a variation in the number of genes it contains within its chromosomes, i.e. some totally missing, and two of others when only one is required. Such problems explain the reason why, where interspecies hybrids occur they are often sterile.

If the number of chromosomes differed between the parents, the problem becomes even greater because this would prevent the "pairing up" from occurring before crossing over took place. In fact, this is typically not a problem with the cat family where the vast majority have 38 chromosomes, and only a few species have 36 chromosomes. There has been one hybrid produced that appears to overcome this problem, but this will be considered later.

All the known mammalian hybrids are thought to be between very closely related species, and virtually all of them have occurred in captivity. The reasons why they do not occur in the wild are really a very straightforward issue of ecology. In the natural world when two very similar species with very similar life styles exist in the same area, generally one will become extinct. This is known as "competitive exclusion", and simply occurs because one species is more efficient at living in the niche available than the other species. A typical example is red squirrels and grey squirrels. Where red squirrels exist and grey squirrels are introduced, the greys take over and simply out compete the reds. A similar type of exclusion is likely to occur with carnivores, hence lions and tigers do not occur in the same range. Lions and leopards do overlap, but the type of prey each takes in the overlap area tends to differ, similarly in the areas where tigers and leopards both exist they tend to prey on different animals. Even where the animals do exist together there is a tendency for behavioural differences to keep them physically apart as individuals.

The situation in captivity is of course very different, individual animals of different may be placed in close proximity to each other without having to compete for food. These animals may have been hand reared with little contact with their own species, and therefore with little learned behaviour from their own parents. What animal contact they had may have been with other species. The further back in time we go the more likely it is that the animal was caught after its parents were killed in a hunt.

Classically, the earliest records of hybrid cats that are quoted are those of the lion/tiger hybrids bred by Mr. Atkins, a travelling menagerist, in the 1820s (Jennison, G. 1927, 56). Atkins probably first produced these hybrids in 1824 when three cubs were born at Windsor. Certainly when he visited Exeter in February 1829, he advertised "*the Lion-Tigers, together in the same den with their Sire and Dam, the male lion and female tigress..*" Such animals are now called ligers, the reverse cross, male tiger and lioness is called a tigon. In fact, this hybridisation may have occurred earlier, before Mr. Atkins' menagerie existed, because the *Exeter Flying Post* for the 31st of July 1788 records a menagerie in Kingsbridge exhibiting a "Lion-Tyger" amongst other animals.

The lion/tiger crosses are somewhat intermediate in appearance and with reduced fertility, it is generally assumed that the males are sterile, but some of the females have proved to be fertile when mated to a (pure-bred) lion or tiger. The liger, though, tends to be large than either of the parents, and one individual currently alive in the United States is said to weigh between 800 and 1200 pounds. This animal is called "Hobbs" and lives at the Sierra Safari Zoo in Nevada. Pictures of him may be found on the Internet easily.

Generally producing hybrid cats is frowned upon in the respectable zoo world, but still occurs in the United States in the less reputable zoos, and on occasions to supply the pet market, because the legal restrictions on exotic pets in some states do not recognise hybrids. At present it is unlikely that there are any lion-tiger hybrids in the UK, possibly the most recent one to be exhibited here would be "Maude" at BelleVue Zoo (Manchester) in the 1940s. Some of the European zoos have a more liberal attitude about hybrids, and I am told that the Tierpark at Hamm in Germany currently has a jaguar x leopard. This is not a novel cross, being recorded as early as 1863. Indeed Flower in 1929 also records a cross between a male lion and a female jaguar x leopard hybrid, which was exhibited at London Zoo after being imported from the United States. The jaguar-leopard parent was probably originally born in Chicago Zoo (Guggisberg, 1975, 140).

Lion-leopard hybrids are known as "leopons" and have been created by mating a male leopard with a lioness at the Hanshin Zoo in Japan. Litters were born in 1959 and 1961, and are illustrated in "*Cats of the World*" by Armand Denis (1964).

A puma-leopard hybrid has been recorded as having been born to a female leopard at the Carl Hagenback Tierpark (Germany) in about 1900, and successfully reared by a fox terrier bitch. It was suggested that this was the sole survivor of 3 sets of twins, and there is no indication of whether it was fertile or not (Gray, 1954, 16).

Hybridisation of the smaller cats appears to have occurred in two very distinct series of circumstances, in the wild, unintentionally and in captivity, very intentionally. The "wild situation" in the UK is thought to have led to the creation of the Kellas cat - a black cat, slightly larger than the domestic cat. This is thought to be a cross between the domestic cat and the European (Scottish) wildcat.

The mechanism by which it occurred is not clear, but it has been suggested that it may have resulted from farm cats crossing with wildcats after going feral when farms were abandoned during the First World War. These animals differ from some of the hybrids between wildcats and domestic cats recorded in captivity, which seem to have the appearance of wildcats (Guggisberg, 1975, 31). However, it is generally thought that the black colour variety of the domestic cat has an advantage in surviving in the wild when it has gone feral, and it is possible that the Kellas may have resulted from black (domestic) farm cats mating with the European wildcat. The offspring of such a cross interbreeding within their own type may have produced the typical Kellas appearance.

The "man-made" hybrids present a very different type of animal that has been created almost entirely for the exotic pet market, typically in America. Possibly the first to be bred intentionally in recent times was the Bengal or Bengali, a cross between the Indian leopard cat and the domestic cat, the objective being to breed an animal with the spots of the leopard cat and the temperament of the domestic cat. This was probably followed next by the "Chaussie" a cross between the jungle cat *(Felis chaus)* and the domestic cat. This animal is currently available from a number of breeders in the United States of America, but one or more may have occurred in the UK.

In 1989 a dead jungle cat was found near Ludlow in Shropshire (*Sunday Express* - 19th February 1989). This animal was subsequently mounted and came into the possession of Karl Shuker. Karl later had his attention drawn to a large a farm cat named Jasper who had the appearance of being a hybrid. Jasper appeared in the appropriate area at the correct time and had the right appearance and size. Unfortunately no DNA work was done to establish his parentage, as it was just a little too early for this.

Once the Bengali and the Chaussie became available it was clearly of major economic sense to try and produce other domestic hybrids for the exotic pet market. This was done mainly in the United States, where such hybrids were often means of getting round the wildlife control laws, and where the money was available for the "creation" of unique and designer pets. The various hybrids that have been "created" are listed each year on the website of the Long Island Ocelot Club. Those listed for 2000 include the savannah, (a domestic cat/serval cross), caraval (caracal/serval cross) [elsewhere this one is sometimes called a "maradi"], and the safari (Geoffroys cat/domestic cat cross). This latter one must be sterile, and is a very strange cross because the Geoffroys cat has 36 chromosomes and the domestic cat 38 chromosomes. Technically this means that the offspring would probably have 37 chromosomes and produce eggs and sperm with different numbers of chromosomes that are most unlikely to function normally. In practice this would make the animals sterile.

The number of captive hybrids that have been recorded within the cat family does tend to indicate that they, as a group, are capable of interbreeding, and possibly more so than most other families within the mammals. The fact that at least one hybrid, the Kellas cat, and maybe a second, the "Ludlow Chaussie", appear to have been produced within the wild in the United Kingdom suggest that through introductions there may be a potential for further hybrids to be created in the wild. This would depend on the introduction of appropriate species to hybridise though. As the domestic cat is one of the commonest pets, has the ability to roam freely, and indeed, to become feral this offers a particular facility for hybridisation.

Many of the reports of sightings of "ABCs" in the United Kingdom are of, what the witnesses claim to be a "black puma". Whilst such animals have been recorded in the wild very rarely, none have been reported as being imported into Europe. Accordingly the chance of such animals having been imported into the United Kingdom and released seems remotely small. The possibility of both a black leopard and a puma that had been released into the United Kingdom meeting up and mating to produce a hybrid cub or cubs would also seem remote.

Even if it did happen it is not certain that the hybrid cats produced would be black, no records of a black leopard (panther) crossed with a puma have ever been recorded. Although it is possible that such (parent) animals may have been released together. Indeed Leslie Maiden, a former lion tamer, has claimed to have released both "a panther and a cougar" (leopard and puma) in Derbyshire in the 1970s (*The Times* 29.02.00).

Even if such a pair of potentially breeding cats were released the chance of these animals successfully producing and rearing a cub or cubs must be very small. On the recorded occasions of a puma x leopard mating having taken place it appears that only one kitten was reared out of six, and that was not reared by the parents. The idea that the reported "black pumas" may be puma-black leopard hybrids seems remotely small, although at least theoretically possible. Whether such hybrids could successfully breed, either amongst themselves, or with either parent species remains yet to be established.

References

- Flower, S.S. (1929) List of the vertebrated animals exhibited in the gardens of the Zoological Society of London 1828 - 1927.1 Mammals. The Zoological Society of London, London.
- Gray, Annie (1954) *Mammalian Hybrids* Commonwealth Agricultural Bureaux, Farnham Royal, Bucks.
- Guggisberg, C.A.W. (1975) Wild Cats of the World David & Charles, Newton Abbot.
- Jennison, G. (1927) Natural History - Animals - An illustrated who's who of the animal world. A & C Black, London.

British Big Fish Records
by
Neil Arnold

EDITOR'S NOTE: Many thanks to Neil for this invaluable contribution to our ongoing *Big Fish Project*.

This article incorporates the CFZ Big Fish Project - an idea concerning all manner of gargantuan sea and lake fish, known or otherwise. For me it is a chance to officially recognise Britain's whoppers and big tiddlers. Basically, the ones that never got away.

For me angling holds a cosy and mysterious energy. As a child I often sat patiently over my still float whilst my father Ron reeled in another tiny gudgeon or flashing bream. The expectancy, even fishing at a small lake, is a magical feeling even if I spent much of my time dissecting maggots. A cruel form of torture that would later come back to haunt me because I can't go near the blighters now!

Every species of freshwater and sea fish has a monster recognised record within its species. There are those too that frustrate anglers as they tell their grandchildren about the one that got away, and there are those true monsters which fishermen dispose of as they fear for their lives on some stormy night.

The following lengthy record covers every record fish caught in Britain and is up to date as of August 2001. Listed are fish species, anglers' names, place and date of catch and the weight which are listed as pounds, ounces and grams

Freshwater/Game Fish:

Fish:	Angler	Location	Date	Weight
BARBEL	Tony Gibson	Great Ouse River, Bucks	2001	19-00-00
BITTERLING	D. Flack	Barway Lake, Cambridgeshire	1998	21 grms
BLEAK	D. Flack	River Lark, Cambridgeshire	1998	00-04-09
BREAM	Kerry Walker	Lodge Lake, Norfolk	2001	18-17-00
BREAM (silver)	E. Flack	Grime Spring, Suffolk	1988	00-15-00
BULLHEAD (miller's thumb)	R. Johnson	B&S Green River, Surrey	1983	00-01-00
CARP	Terry Glebioska	Conningbrook, Kent	2001	59-07-00
CARP (crucian)	Adrian Eves	Surrey Pit	1999	04-05-08
CARP (grass)	S. Lavin	Church Lake, Berkshire	1999	33-12-00
CATFISH (black, bullhead)	Bobby Barnes (age 14)	Lake Meadows, Essex	1999	01-00-01
CATFISH (wels)	R. Garner	Withy Pool, Henlow, Beds	1997	62-00-00
CHUB	Smith	River Tees, Co. Durham	1994	08-10-00
CHARR (arctic) GAME FISH NATURAL	W. Fairbairn	Loch Arkaig, Scotland	1995	09-08-00
DACE	J.L. Gasson	Little Ouse, Norfolk	1960	01-04-04
EEL	Master S. Terry	Kingfisher Lake, Hants	1978	11-02-00

GOLDFISH (brown)	D. Lewis	Surrey Still Water Pond	1994	05-11-08
GRAYLING GAME FISH NATURAL	S.R. Lanigan	River Frome, Dorset	1989	04-03-00
GUDGEON	H. Hull	River Nadder, Wiltshire	1990	00-05-00
MINNOW	J. Sawyer	Whitworth Lake, Spennymoor	1998	00-00-13.5
ORFE (golden)	Michael Wilkinson	Lymm Lake, Cheshire	2000	08-05-04
PERCH	J. Shayler	A private lake, Kent	1985	05-09-00
PIKE	R. Lewis	Llandegfedd, Wales	1992	46-13-00
PUMPKINSEED	D.L. Wallis	Whessoe Pond, Co. Durham	1987	00-04-09
ROACH	R.N. Clarke	Dorset Stour	1990	04-03-00
RUDD	E.C. Alston	Mere Nr. Thetford, Suffolk	1933	04-08-00
RUFFE	R.J. Jenkins	West View Farm, Cumbria	1980	00-05-04
SALMON (atlantic) GAME FISH NATURAL	Miss G.W. Ballatine	River Tay, Scotland	1922	64-00-00
SCHELLY (skelly)	S.M. Barrie	Haweswater Reservoir, Cumbria	1986	02-01-09
STICKLEBACK (3-spined)	D. Flack	High Flyer Lake, Cambs	1998	7 grms
TENCH	Darren Ward	Southern Stillwater	2001	15-00-00
TROUT (american brook char) GAME FISH CULTIVATED	E. Holland	Frontburn Resv. Northumberland	1998	08-03-00
TROUT (brown) GAME FISH NATURAL & CULTIVATED	A. Finlay	Loch Awe, Scotland	1996	25-05-12
TROUT (rainbow) GAME FISH CULTIVATED	C. White	Dever Springs, Hants (Fishery)	1995	36-14-08

TROUT (sea)
GAME FISH
NATURAL   J. Farrent   Calshot Spit,
                                      River Test   1992   28-05-04

WALLEYE
(pikeperch)   F. Adams   The delph, Norfolk   1934   11-12-00
ZANDER (pikeperch)
                   D. Lavendar   Fen Water, Cambs   1998   19-05-08

BOAT RECORDS: (B)
SHORE RECORDS☺S)

ANGLER FISH
(B)   S.M.A. Neill   Belfast Lough,
                                      N. Ireland   1985   94-12-04
(S)   H. Legerton   Canvey Island, Essex   1967   68-02-00

BASS
(B)   P. McEwan   Reculver, Herne Bay,
                                      Kent   1987   19-09-02
(S)   D.L. Bourne   Southern breakwater,
                                      Dover, Kent   1988   19-00-00
BLACK FISH
(B)   J. Semple   Off Heads of Ayr,
                                      Scotland   1972   03-10-08
(S)   Ostler   Aldbrough Beach,
                                      East Yorks   1998   05-14-08
BLUEMOUTH
(B)   Anne Lyngholm   Loch Shell, Scotland   1976   03-02-08
Qualifying Weight   01-00-00

BOGUE
(B)   K. McBride   Eddystone Reef   1981   01-13-00
(S)   S.G. Torode   Pembroke Guernsey,
                                      Channel Isl   1978   01-15-04

BONITO

(B)   J. Parnell   Torbay, South Devon   1969   08-13-04
(S)   P. Blanning   St. Brides Way,
                                      Pembroke, Wales   1996   02-10-05
BREAM (black)

(B)	J.A. Garlick	From wreck, off South Devon coast	1977	06-14-04
(S)	M. Guilmoto	Cachaliere Pier, Channel Islands	1994	05-00-02

BREAM (couch's sea)
(B)	A. Drysdale	Newquay, Cornwall	1998	04-12-09
(S)	M. Thomson	St. Peter Port, Channel Islands	1996	02-00-07

BREAM (gilthead)
(B)	C.J. Bradford	Salcombe Estuary, Devon	1991	09-15-08
(S)	Colin Carr (age 14)	Salcombe Estuary, Devon	1995	10 05 08

BREAM (pandora)
	C. Stone	Newquay, Cornwall	1997	03-06-12

BREAM (ray's)
(B)	J. Holland	Barra Head, Hebrides	1978	06-03-13
(S)	G. Walker	Crimdon Beach, Hartlepool	1967	07-15-12

BREAM (red)
(B)	B. Reynolds	from wreck, Mevagissey, Cornwall	1974	09-08-12
(S)	A. Salmon	Alderney Lighthouse, Channel Isl	1979	04-07-00

BRILL
(B)	A. Fisher	Derby Haven, Isle Of Man	1950	16-00-00
(S)	B.M.K. Fletcher	Nr. Vason Bay, Guernsey	1980	07-07-08

BULL HUSS
(B)	M.L. Hall	Minehead, Somerset	1986	22-04-00
(S)	G. Ebbs	Pwllheli Beach, Gwynedd, Wales	1992	19-14-00

CATFISH
(B)	S.P. Ward	Five miles off Whitby, Yorks	1989	26-04-00
(S)	G.M. Taylor	Stonehaven, Scotland	1978	12-12-08

COALFISH
(B)	D. Brown	wreck, South of Eddystone	1986	37-05-00
(S)	M. Cammish	Filey Brigg, Yorkshire	1995	24-11-12

COD
(B)	Noel Cook	North Sea,		

			Whitby, N. Yorks	1992	37-05-00
(S)		B. Jones	Tom's Point, Barry, Wales	1966	44-08-00

COMBER

		Master B. Phillips (age 10) Mounts Bay, Penzance, Cornwall	1977	01-13-00
		Qualifying Weight		00-12-00

CONGER

(B)	V. Evans	Wreck, Berry Head, South Devon	1995	133-04-00
(S)	M. Larkin	Devil's Point, Plymouth	1992	68-08-00

DAB

(B) Master R. Islip		Gairloch, Wester Ross Scotland	1975	02-12-04
(S)	M.L. Watts	Morfa Beach, Port Talbot, Wales	1936	02-09-08

DOGFISH (black mouthed)

J.H. Anderson		N.W. Poll Point, Loch Fyne, Scotland	1977	02-13-08
		Qualifying Weight		00-12-00

DOGFISH (lesser spotted)

(B)	G. Griffiths	Port Logan, Galloway, Scotland	1994	04-06-08
(S)	S. Ramsey	Abbey Burnfoot, Scotland	1988	04-15-03

FLOUNDER

(B)	A.G.L. Cobbledick	Fowey, Cornwall	1956	05-11-08
(S)	B. Sokell	River Teign, South Devon	1994	05-07-00

FORKBEARD (greater)

Miss M. Woodgate	Falmouth Bay, Cornwall	1969	04-11-04
--	Qualifying Weight		01-00-00

GARFISH

(B)	A. Saunders	Mounts Bay, Penzance, Cornwall	1994	03-09-08
(S)	F. Williams	Porthoustock, South West Cornwall	1995	03-04-12

GREATER WEEVER

(B)	T. Griffiths	Mounts Bay, Penzance, Cornwall	1994	02-00-13
(S)	B. Bawden	Porthcurno, Cornwall	1989	01-07-15

GURNARD (red)
(B)	Benjamin Le Noury (age 14)	Havre Gosselin, Sark, Channel Islands	1995	02-13-11
(S)	D. Johns	Helford River, Glebe Cove, Cornwall	1976	02-10-11

GURNARD (grey)
(B)	D. Swinbanks	Caliach Point, Mull, Scotland	1976	02-07-00
(S)	P. El-Balawi	Newquay Headland, Cornwall	1989	01-10-08

EDITOR'S NOTE: Never been one to avoid jumping to quasi fortean conclusions, it is mildly amusing, I think, to point out the surrealchemical properties of the gurnard records. It seems likely that 'Glebe Cove' on the Helford River where the record breaking red gurnard was caught is none other than 'Grebe Beach' where `Doc` Shiels saw Morgawr in 1976. It is also worth pointing out that the Caliach is the hag aspect of the faced Pagan Goddess. (The grey gurnard record is from Caliach point!)

GURNARD (streaked)
(B)	M. Adams	Poole Bay, Dorset	1994	01-02-00
(S)	H. Livingstone Smith	Loch Goil, Firth of Clyde, Scotland	1971	01-06-08

GURNARD (yellow or tubfish)
(B)	C.W. King	Wallasey, Cheshire	1952	11-07-04
(S)	G.J. Reynolds	Langlan Bay, Wales	1976	12-03-00

HADDOCK
(B)	G. Bones	Off Falmouth, Cornwall	1978	13-11-04
(S)	G.B. Stevenson	Loch Goil, Scotland	1976	06-12-00

HADDOCK (norway)
(B)	T. Barret	Thames Estuary, Southend on Sea	1975	01-13-08
(S)	F.P. Fawke	Southend Pier, Essex	1973	01-03-00

HAKE
(B)	R. Roberts	Loch Etive, Scotland	1997	25-12-14
(S)	W.S. Parry	Morfa Beach, Port Talbot, Wales	1984	03-08-02

HALIBUT
(B)	C. Booth	Dunnet Head, Scrabster, Scotland	1979	234-00-00

Qualifying Weight   10-00-00

<u>Editor's Note:</u> As we were preparing this volume a record breaking halibut weighing 20 st (127kg), was caught off Rockall, on a trawler by James Lovie on April 17th 2001. The fish was landed at Kinlochbervie, on the west coast of Sutherland.

HERRING
Master. Barden (age 13)	off Bexhill-on-Sea, Essex	1973	01-01-00
--	Qualifying Weight		01-00-00

JOHN DORY (St. Peters fish)
J. Johnson	off Newhaven, East Sussex	1977	11-14-00
--	Qualifying Weight		03-00-00

LING
(B)	J. Webster	off Bridlington, Yorks	1989	59-08-00
(S)	K. Smith	Seasons Point, Plymouth	1994	21-10-00

LUMPSUCKER
(B)	J. McCoy	Tyne Estuary	1986	10-12-10
(S)	Andrew Perry (age 15)	Weymouth, Dorset	1987	20-09-12

MACKEREL
(B)	W.J. Chapple	Penberth Cove, West Cornwall	1984	06-02-07
(S)	M.A. Kemp	Berry Head Quarry, Devon	1982	05-11-14

MACKEREL (spanish)
(B)	C.J. Bradford	Chattern Reef, Salcombe, Devon	1995	01-02-00
(S)	G. Scammell	Lamorna Cove, Penzance, Cornwall	1995	01-05-07

MEGRIM
(B) Master P. Christie (age 15)		Gairloch, Scotland	1973	03-12-08
(S)	F.J. Williams	Porthcurno, Cornwall	1987	02-06-01

MONKFISH
(B)	C.G. Chalk	South of Shoreham, Sussex	1965	66-00-00
(S)	G.S. Bishop	Llwyngwril Beach, N. Wales	1984	52-04-00

MULLET (golden grey)

(B)	J. Case	Brixham Harbour, South Devon	1999	02-13-06
(S)	J. Reeves	Old Man's Nose, Alderney, Channel Isl	1991	03-00-04
Also = D. Heward		Christchurch Harbour, Dorset	1994	03-00-04

MULLET (red)

(B)	M.T. Hamel	Petils Bay, Bordeaux Vale, Guernsey	1981	03-07-00
(S)	A. Wright	Longy Bay, Alderney, Channel Islands	1996	03-15-00

MULLET (thick lipped grey)

(B)	P.C. Libby	Portland, Dorset	1952	10-01-00
(S)	R.S. Gifford	The Leys, Aberthaw, Glam. Wales	1979	14-02-12

MULLET (thin lipped grey)

(B)	Mrs Ann Copley	River Medway, Kent	1991	05-15-00
(S)	N. Mableson	Saltside, Oulton Broad, Suffolk	1991	07-00-00

OPAH

A.R. Blewett	Mounts Bay, Penzance, Cornwall	1973	128-00-00
--	Qualifying Weight		20-00-00

PELAMID (honito)

(B)	J. Parnell	Torbay, South Devon	1969	08-13-04
(S)	P. Blanning	St. Brides Bay, Pembroke	1996	02-10-05

PERCH (dusky)

D. Cope	off Duriston Head, Dorset	1973	28-00-00
--	Qualifying Weight		07-00-00

PLAICE

(B)	H. Gardiner (age 16)	Longa Sound, Scotland	1974	10-03-08
(S)	R. Moore	Southbourne Beach, Bournemouth	1989	08-06-14

POLLACK

(B)	W.S. Mayes	Dungeness, Kent	1987	29-04-00
(S)	C. Lowe	Abbotsbury Beach, Dorset	1986	18-04-00

POUTING

(B)	R. Armstrong	off Berry Head, Devon	1969	05-08-00
(S)	R. Andrews	Pembroke, Guernsey, Channel Isl	1991	04-09-00

PUFFER FISH

--	Qualifying Weight		04-

00-00				
S. Atkinson	Abottsbury, Dorset		1985	06-09-04

RAY (bottle-nosed)

R. Bulpitt	off The Needles, Isle Of Wight		1970	76-00-00
- -	Qualifying Weight			15-00-00

RAY (blonde)

(B)	Ian Dobson	Nab Tower, off Isle Of Wight	2000	38-09-00
(S)	C.M. Reeves	Mannez Targets, Alderney	1986	32-08-00
ALSO =	K. Frain	Grosnez, Jersey North coast	1994	32-08-00

RAY (cuckoo)

(B)	V. Morrison	Causeway Coast, Northern Ireland	1975	05-11-00
(S)	C. Wills	North Cliffs, Cornwall	1981	04-10-00

RAY (eagle)

M.A. Drew	Nab Waters, S.E. of Colver, Isle Of Wight	1989	61-08-00
- -	Qualifying Weight		15-00-00

RAY (electric)

(B)	N.J. Crowley	Dodman Point, Cornwall	1975	96-01-00
(S) Master M.A. Willis		Porthallow, Cornwall	1980	52-11-00

RAY (marbled, electric)

(B)	B.T. Maguir	St Aubins Bay, Jersey, Channel Isl	1983	02-08-08
(S)	M.E. Porter	North coast of Jersey, Channel Isl	1990	13-15-11

RAY (sandy

	Qualifying Weight		02-00-00
- -	Qualifying Weight		02-00-00

RAY (small eyed)

(B)	Mrs Susan Storey	Watchet Harbour, N. Somerset Coast	1991	17-08-00
(S)	N.J. Wood	South Wales	1991	15-00-08

RAY (spotted)

(B)	A. Hodge	Whitsands Bay, Cornwall	1998	08-10-08
(S)	G.D. Bowen	Mewslade Bay, South Wales	1980	08-05-00

RAY (sting)
(B)	P Burgess	River Blackwater, Essex	1996	72-02-00
(S)	K. Wyatt	Fairbourne Beach, Gwynedd, S. Wales	1991	54-09-06

RAY (thornback)
(B)	J. Wright	Liverpool Bay	1981	31-07-00
(S)	S. Ramsay	The Ross, Kirkcudbright, Scotland	1985	21-12-00

RAY (undulate)
(B)	S. Titt	4-miles south St. Aldhelms, Dorset	1987	21-04-08
(S)	K. Skinner	St. Catherines Lighthouse, Jersey	1983	21-04-00

ROCKLING (3-bearded)
(B)	C. Hurst	South of the Isle Of Wight	1992	03-04-04
(S)	W. Crothers	Holborn Head, Scotland	1992	03-01-12

ROCKLING (shore)
- -		Qualifying Weight		00-12-00
	D. Lane	Chesil Cove, Dorset	1992	01-09-12

SALMON (coho)
Qualifying Weight			01-00-00
R.J. McClaren	Petit Port, Guernsey, Channel Islands	1977	01-08-01

SCAD (horse mackerel)
(B)	M.A. Atkins	Torbay, South Devon	1978	03-05-03
(S)	R. Dillon	Cockenzie Power Station, Scotland	1981	03-00-14

SEA SCORPION (short spined)
(B)	R. Stephenson	Keppell Pier, Great Cumbrae Isle, S'land	1973	02-03-00
(S)	B. Logan	Whitley Bay, Tyne & Wear	1982	02-07-08

SHAD (allis)
- -		Qualifying Weight		02-00-00
P.B. Gerrard		Chesil Beach, Dorset	1977	04-12-07

SHAD (twaite)
(B)	D. Protheroe	Barry, South Wales	1985	02-04-02
(S)	J. Martin	Garlieston, Wigtownshire, Scotland (S.W.)	1978	02-12-00

SHARK (blue)
N. Sutcliffe	Looe, Cornwall	1959	218-00-00
- -	Qualifying Weight		40-00-00

SHARK (mako)
Mrs J.M. Yallop	North West of Eddystone Light	1971	500-00-00
- -	Qualifying weight		40-00-00

SHARK (porbeagle)
C. Bennet	Dunnet Head, Scotland	1993	507-00-00
- -	Qualifying Weight		40-00-00

SHARK (six-gilled)
F.E. Beeton	Penlee Point, Plymouth, Devon	1976	09-08-00
- -	Qualifying Weight		05-00-00

SHARK (thresher)
S. Mills	South Nab Tower, Portsmouth, Hants	1982	323-00-00
- -	Qualifying Weight		40-00-00

SKATE (common)
(B)	R. Banks	off Tobermory, Inner Hebrides, Scotland	1986	227-00-00
(S)	G. MacKenzie	Breasclete Pie Loch Ruag, Isle of Lewis	1994	169-06-00

SMOOTHOUND
(B)	A.T. Chilvers	Heacham, Norfolk	1969	28-00-00
(S)	D. Ellis	Hillhead Wall, Hants	1997	19-08-06

SMOOTHOUND (starry)
(B)	A. Bowering	off Minehead, Bristol Channel	1998	28-02-00
(S)	D. Carpenter	Beach at Bradwell-on-sea, Essex	1972	23-02-00

SOLE
(B)	M. Eppelein	Great Bank, Guernsey	1993	04-01-12
(S)	N.V. Guilmoto	South Coast Boulders, Alderney	1991	06-08-10

SOLE (lemon)
(B)	J. Gordon	Loch Goil Head, Firth of Clyde, Scotland	1976	02-02-00
(S)	W.N. Callister	Victoria Pier, Douglas, Isle Of man	1980	02-07-11

SPURDOG
(B)	P. Barrett	off Porthleven, Cornwall	1977	21-03-07

(S)	R. Legg	Chesil Beach, Dorset	1964	16-12-08

SUNFISH
(B)	T. Sisson	off Saundersfoot, Wales	1976	108-00-00
(S)	M. Merry	Fisherman's Cove, North Cliffs, Cornwall	1976	49-04-00

TADPOLE-FISH
(B)	S. Bishop	off Christchurch, Dorset	1991	01-04-04
(S)	N. Conn	Seaham Beach, Co. Durham	1990	01-05-12

TOPE
(B)	R. Chatfield	off Bradwell-on-sea, Essex	1991	82-08-00
(S)	R.J. White	Baggy Point, North Devon	1982	58-02-00

TORSK
D. MacKay	Pentland Firth, Scotland	1982	15-07-02
- -	Qualifying Weight		03-00-00

TRIGGERFISH
(B)	P. Robinson	Langland, Swansea	1995	05-08-11
(S)	R. Lovering	Woody Bay, North Devon	1995	05-08-11
ALSO=	K. Lydiard	Lynmouth Rocks, North Devon	1995	05-14-08

TUNNY
L. Mitchell-Henry	Whitby, Yorks	1933	851-00-00
- -	Qualifying Weight		40-00-00

TUNNY (big eyed)
- -	Qualifying Weight		30-00-00
A. Pascoe (age 15)	Newlyn Harbour, Devon	1985	66-12-00

TUNNY (long-finned)
B. Cater	Salcombe, Devon	1990	04-12-00
- -	VACANT		

TURBOT
(B)	R. Simcox	Salcombe, Devon	1980	33-12-00
(S)	J. Dorling	Dunwich Beach, East Suffolk	1973	28-08-00

WHITING
(B)	N.R. Croft	Manacle rocks, Falmouth Bay, Cornwall	1981	06-12-00
(S)	T. Dell	Abbotsbury, Dorset	1984	04-00-07

WHITING (blue)
J. Anderson		Loch Fyne, Scotland	1977	01-12-00
- -		Qualifying Weight		00-12-00

WITCH
- -		Qualifying Weight		00-12-00
T. Barathy		Colwyn Bay, N. Wales	1967	01-02-13

WRASSE (cuckoo)
(B)	K. Marden	Dungeness, Kent	1998	02-07-12
(S)	M. Hooper	Rosaire, Herm, Channel Islands	1999	01-12-09

WRECKFISH
B. McNamara	South of Eddystone	1980	10-10-00
- -	Qualifying Weight		01-00-00

ll new record fish caught by rod should be registered with the BRITISH RECORD FISH COMMITTEE. Rules and regulations can be found on the *Angling Times* website.

The Committee exists to recognise and publish record weights of both fresh and saltwater fish caught on rod and line by fair angling methods in the waters of England, Wales, Scotland, Northern Ireland and the Channel Islands. Its aims are:

1) To provide an adjudicating body to which marine and freshwater anglers may submit claims for record fish taken by fair rod and line angling.
2) To investigate all such record claims to the fullest possible extent and maintain a permanent record of such investigations.
3) To establish and maintain accurately a list of British fish, marine and freshwater, to record size and to publish the list frequently and make it readily available to all interested persons.

Procedures, methods of capture, weight, identification of species, protected fishes and other details available from: BRFC, 51A Queen Street, Newton Abbot, Devon, TQ12 2QI

Additional Records:

TROUT (rainbow) GAME FISH   RESIDENT
J. Hammond   Hanningfield Reservoir, Essex   1998   24-01-04
TROUT (rainbow)

GAME FISH WILD   VACANT

WRASSE (ballan)
(B)   A. Heart   off Jersey, Channel Islands   1999   09-07-12
(S)   P. Hegg   Portland, Dorset   1998   09-01-00

The Fortean Zoological aspects of the *Biggles* books of Captain W.E Johns

by Graham Inglis

Captain W. E. Johns' writing, like that of detective-writer Agatha Christie, generally adhered to a prosaic puzzle-solving theme - while, very occasionally, his *Biggles* stories would delve into science fiction type avenues. Christie's "*The Big Four*" was probably her most out-and-out SF dalliance - but generally, her occasional excursions into the paranormal dealt with supernatural themes such as seánces. Meanwhile, James Bigglesworth ("Biggles"), best known for shooting marauding Germans out of British skies during two world wars, occasionally found himself in the relatively-unfamiliar territory of grappling with bizarre cryptids such as 8-foot crabs, or even facing electrically-controlled centipedes in Tibet.

This is a review of some of Biggles' cryptid encounters, with zoological comments by Richard Freeman.

Background

The author William Earl Johns was born in 1893 in Hertfordshire, England, the son of a tailor. During World War I he served as machine gunner with the Royal Flying Corps, the precursor of the RAF and in early 1918 he finally realized his ambition to become a fighter pilot in France. However he was soon shot down over enemy territory and captured. He managed to escape, but was caught again and sentenced to death - but the November Armistice came to the rescue, thus preserving one of the most successful and prolific writers of the 20th Century.

He remained with the Royal Air Force until 1931, as a flight instructor and later as a recruiting officer, finally reaching rank of Flying Officer. His famous hero, Biggles, first appeared in print in the early 1930s. Johns drew heavily on his own war experiences and on those of his friends, and Johns eventually was to write more than 90 books featuring Biggles before his death in 1968.

In a sharp departure from the third person narrative of the normal Biggles stories, Biggles Charter Pilot is a series of tales narrated by Ginger in idle moments during the Second World War, when he and other pilots of his patrol were on standby. He reminisces about a time before the war when Biggles placed an advertisement in the newspapers offering his services as a charter pilot. Their main client was an archetypal absent-minded biology professor called Dr. Augustus Duck, who was to enjoy a succession of cryptozoological encounters.

Biggles - Charter Pilot: The Inquisitive Dodos

Following a remark from a pilot that "chivalry in war is as dead as the dodo," a discussion about the dodo prompts Ginger to describe a Dr. Duck-inspired search for the dodo.

'[Duck] *had a feeling that, considering the thousands of uninhabited islands in the Indian Ocean, there was a chance of a few dodos remaining, hidden away on some lonely isle. That wasn't a question that could be answered by sitting at home in an arm-chair, guessing. Having studied the atlas, at no small expense he decided to make a protracted cruise over a few dozen islands which he had selected as the most likely spots to find a surviving dodo...*'

Ginger leaned back on the settee, put his feet on a chair, and continued:

'*One day we were making a fairly long hop back to Madagascar to refuel, when the weather turned nasty. I could see Biggles was getting worried. Suddenly an island showed up on the horizon. I call it an island, but islet would probably be a more accurate description. It wasn't much more than a mile long, two or three hundred yards wide, and. not much more than ten or twelve feet above sea level at the highest point.*'

Ginger describes how, after they landed the amphibious aircraft on the sea and then taxied onto dry ground, fog closed in and they were stranded for the night. At daybreak the fog cleared, revealing an astonishing sight.

'*We stared, and kept on staring. For covering the entire islet, from end to end, were thousands and thousands of white sea-birds, great ridiculous-looking creatures with enormous beaks and curly tails. There they stood, all staring at us, so thick that a jenny-wren would have found it hard to find a place to perch. You should have seen Biggles's face.*'

"Unless I'm dreaming," he said, "they're dodos. We couldn't find them, but it seems that they've found us. Wake Donald [Dr. Duck]. Don't make a noise or you may scare the birds *away before he can feast his eyes on them."*

'I went into the cabin and woke the others. They followed me out, Donald nearly beside himself with excitement. Out came his notebook, and we watched him in his element for about twenty minutes. Biggles then pointed out that he was sorry to disturb the party, but as the sea was rising fast we ought to be on our way before we were washed away.

'Biggles waved his arms, and. addressing the flock, told them to push off.
'The birds utterly ignored him. They just stood there like a lot of dignified old ladies, looking at us with expressionless faces. We shouted. We jumped, we yelled, but still they didn't move a step or flap a wing. Then we remembered that they couldn't fly. Biggles went into the cabin, and returning with a gun, blazed away over their heads. They didn't bat an eyelid.

'Algy roared with laughter, but Biggles, after a glance at the sea, stopped him. "This isn't in the least funny," he said seriously. "If those birds refuse to move we are very soon going to be in a jam. Look at those waves! We've got to clear a runway, and we've no time to lose. Come on, get busy."

'I got hold of a bird and started pushing, but another simply walked into its place. We all pushed; we heaved and shoved, but all we did was get ourselves hot and bothered. As fast as we cleared a spot fresh birds just strolled on to it. Algy said, "We shall have to shoot the lot, and throw them into the sea." Biggles answered, "Don't be a fool; we should need a million rounds of ammunition." I could see that he was really worried. "I've been mixed up in some crazy adventures," he said, "but this beats the lot. I'm dashed if I know what to do, and that's a fact." '

Ginger describes how Biggles soon decided to fire up the aeroplane engines to clear a runway.

'We managed to hang on [to the tail unit], but either the noise or the blast of air was too much for the dodos. The slipstream blew them into heaps, like piles of feather cushions. As long as I live I shan't forget that picture - dodos on their backs, sliding on their faces, with their silly little wings stuck out. In a minute or two we had a clear run of a hundred yards or so. It wasn't very wide, but it was wide enough, provided the birds didn't come back before we could get off. The runway was, of course, behind the machine, so we hauled the tail round until the nose was pointing in the right direction. Then we fairly fell into the aircraft...

'We got off by the thickness of a pig's bristle. Looking down, the runway had already disappeared. The islet was once more a solid mass of dodos. Where they'd come from we don't know; where they went, if they went anywhere, we don't know either. Speaking personally, I don't jolly well care. Biggles took a shot at the sun, picked up our course, and a couple of hours later we were in Madagascar, finishing up with a pint of petrol in the tank.'

Ginger got up and stretched. 'Well, that's all,' he said. 'But if anyone ever tells you that the dodo bird is extinct, just refer him to Biggles. What he thinks of those dizzy fowls is nobody's business.'

NOTE BY RICHARD FREEMAN: The dodo was a giant flightless pigeon. There were in fact two species, the Mauritius dodo, *Raphus cucullatus,* with its grey plumage, and the Reunion dodo, *Victoronis imperilas,* with white feathers. Both the islands of Mauritius and Reunion also had races of a bird known as the solitaire, *Pezophaps solitarius* and *Ornithaptero solitaria* respectively. These were more slender than the dodos, with longer necks.

Dodos have been portrayed as fat, slow, stupid birds but recent studies seem to show that they were capable of running quite quickly. Specimens brought to Europe may have been suffering from an inappropriate diet. On their Indian Ocean Island there were no terrestrial predators, hence the loss of flight. A striking example of parallel evolution can be seen in New Zealand's owl parrot, *Strigops habroptilus,* a bird itself facing extinction. The hunting of the dodo and its kin by mariners, coupled with the introduction of goats, dogs, cats, rats, monkeys, and mongoose that altered the delicate Iland ecosystem, forced these birds into extinction by 1669.

Some researchers such as Bill Gibbons claim to have been told stories of encounters with living dodos on remote Mauritian beaches. Sadly there can be little hope of these stories having any basis in fact. Mauritius and Reunion have been described as ecological disaster areas. Almost none of the indigenous forests are left and there is just no room left to hide a flightless turkey sized bird. Roy Mackal has pointed out that many of the thousands of smaller islands in the Indian Ocean remain unexplored. Undiscovered species of dodo *like* birds may inhabit them. Known species of giant pigeon exist such as the crowned pigeon *Goura victoria* of New Guinea so flightless, ground living relatives are not out of the question.

Mackal sites Ile Tromelin north of Madagascar as a possibility. If dodo-related birds do live on these tiny islands they will be much smaller than known species. One wonders what W.E.John`s huge dodos were eating on their speck of an islet!

The Cruise of the Condor

In the inter-War years of the 1920s and 30s, the interiors of Africa and South America had yet to be divided up into self-determining territories. Then, they were still swathes of mostly-unmapped continent, still being explored by the (almost invariably) white men of Britain, Europe and the USA and, once reached, it was a matter of being vigilant against tribal or animal attack. Indeed, in some stories of the time, little if any distinction was made between the two hazards.

However, by the time Biggles visited such areas, inter-racial relations had progressed somewhat. This was the 1930s and offering beads and mirrors to the 'savages' and worrying about large cooking pots was no longer standard operating procedure.

1933, and we find Biggles on the hunt for Inca treasure. Temporarily separated from the rest of his party, he's descending a perilous path set into the side of a raised plateau in the hinterland of Brazil.

"He was about halfway to the bend when a shadow swept across the face of the cliff just in front of him, and, looking round without any particular alarm to ascertain the cause, he saw the largest bird he had ever seen in his life. It was snow white from beak to tail, and he judged it to measure a full twenty feet from wing tip to wing tip. Its cruel curved beak and formidable talons betrayed it to be a bird of prey, and he watched its stately flight in admiration. "I didn't know there were such things as white eagles," he mused as he continued his way.

(At this juncture a footnote by Johns comments: Although Biggles did not know it, he was looking at what is probably the rarest bird in the world, the magnificent king condor of the Andes, the existence of which travellers in the Cordillera have reported from time to time. Several attempts have been made to take one dead or alive, but without success. It was named the king condor because ordinary condors seemed subservient to it.)

He had only taken a few steps when a noise of rushing air made him turn quickly with an unpleasant consciousness of danger. The bird was swooping down on him, and he dropped to his knees just as it swept over him, the long curving talons that would have torn his face to ribbons passing within a foot of his head. He was on his feet the instant it had passed, hurrying towards the bend, for the narrow shelf to which he clung was no place for an encounter with either bird or beast.

But before he had taken six steps it was clear that the bird had no intention of abandoning its presumed prey, for it soared up in a steep climbing turn and then dropped like a stone towards him, pinions raised, talons projecting viciously below. Biggles grabbed in his pocket for the automatic which he had carried ever since the affair of the Indians, but before he could use it the bird was on him. Instinctively he flung himself down at full length as the bird swept past in a vertical bank at the end of its dive, and the rush of air that followed it nearly blew him from the ledge. He jerked up the automatic, and three fingers of flame leapt from the muzzle. Crack-crack-crack! it spat viciously.

He knew instantly that the bird was hard hit. It faltered in its flight, actually dropping a few feet, and then, recovering itself with an effort, flew to a neighbouring crag, where it settled and then collapsed with outstretched wings. Twice it made a stupendous effort to rise, but failed, and finally, after a convulsive flap of its great wings, it lay still.

"Sorry, old bird, but you asked for it," muttered Biggles in a tone of sincere regret as he dropped the automatic back into his pocket, for he was genuinely sorry that he had been forced to destroy such a noble-looking creature.

He had nearly reached the bend when a great noise of rushing wings caused him to look up with a start. The air was full of huge, dark-brown birds falling towards him from out of the blue sky. He saw them land, one after the other, with effortless ease on the rock where the white bird lay; then they rose in a cloud and swept towards him with a directness that left no doubt as to their intentions. He waited for no more. As swiftly as he dared he sped along a pathway where, the day before, he would have hesitated to take a single step. He reached the bend with the revengeful winged subjects of the dead king close behind him, knowing that unless some shelter quickly revealed itself he was lost. A single bird he might, and had indeed, vanquished, but a whole flight was beyond his ability to cope with.

He slowed down as he reached the bend lest his impetus should carry him over the brink, and, turning the corner, saw that the path ... ended abruptly-in mid-air. At his feet lay a broad ravine. Forty feet below, a boiling rapid, swollen by the recent rain, raced with headlong, pent-up fury between its narrow rocky confines.

Biggles knew that he was at the end of his tether, for the birds were already swooping to the attack. There was only one thing to do, and he made up his mind quickly. Backing a few yards up the path for the takeoff, he sped down the slope and launched himself into space. He knew before he jumped that it was too wide for him, but nevertheless he actually reached the opposite bank; for one awful moment he struggled to maintain his balance, but a rock gave way under his weight and he plunged down into the whirling torrent below.

RICHARD FREEMAN'S NOTE: Giant condors did exist and in a rare example of literary understatement they grew larger than the Captain's bird. The teratorns flourished in the Pleistocene epoch 2 million to 10 thousand years ago. The largest of them, *Argentavis magnificens* boasted a wingspan of 25 feet! Teratorns evolved to feed on the carcasses of new world megafauna such as giant sloths, elephants, horses, and camels. With the mass extinctions of these huge animals at the end of the Pleistocene it was assumed the teratorns died out too. This presumption may have been premature.

Many American Indian tribes have a tradition of the thunderbird, a bird so massive that its wing beats caused thunder. Some believed that the thunderbird brought the storms on its wings. These could be dismissed as folktales if it were not for consistent sightings in the southern US involving birds of staggering proportions.

Modern thunderbird sightings suggest birds with wingspans between 10 and 30 feet. Animal deaths and disappearances have been blamed on them as have human disappearances but such allegations remain unproven. Some Indians have postulated that teratorns survived by feeding on the carcasses of the American bison *Bison bison.*

The most dramatic encounter took place on July 25 1977 when ten-year-old Marlon Lowe was attacked by a giant vulture like bird whilst playing in his back yard in Lawndale, Illinois. He was carried more than 30 feet before the bird, one of a pair, dropped him. The attack caused his red hair to turn white.

One problem remained with the teratorn / thunderbird equation: new world vultures are not closely related to old world vultures or indeed to any other birds of prey.

They are more closely related to storks. New world vultures cannot grip with their talons like true raptors and ergo could not grip and lift prey off the ground. Unless we are dealing with radical unknown species or unless we have made serious mistakes in our evaluation of teratorns, these giant

Biggles: Charter Pilot - The Oxidized Grotto

In a moment of relaxation between airborne patrols, the flight crews are discussing airplane construction.

"Talking of metals," remarked Algy Lacey, "it was once my privilege - I can't say pleasure - to see a specimen of what must be the most uncommon metal in the world, a metal so rare that it was lost to science for thousands of years. In fact, it has often been asserted that no such metal exists, or ever did exist."

"Then how did people come to know about it in the first place?" queried Tug Carrington suspiciously.

"Because it is mentioned by more than one ancient scribe."

Pressed for details, Algy refers his interested audience of airmen to Ginger, who, over a cup of tea, describes how Dr. Duck proposed an expedition to Borneo to find a cave in which every living thing was the same color – chalk white. They eventually find the cave, the entrance of which has a coating of dirty white moss, which crunched underfoot. Entering the cave,

...it quickly became paler, and then dead white, snow white, an extraordinary spectacle. And now a singular state of affairs arose. Everything being white, we could see nothing but shadows. I bumped into a white bush without seeing it, and Algy nearly put his foot on a fair-sized snake. It was as white as the floor, so it's not surprising that he didn't see it. The snake, a queer, blunt-headed creature, hissed and went off down the cave. We looked at the bush. The flower, a kind of orchid, was white. The leaves were white and the stalks were white, so you could only tell which was which by the shape. There were flies on it, white flies, while white bats fluttered over our heads. In the light of our torches they looked like enormous white butterflies. You can't imagine how peculiar the whole effect was.

Donald, of course, was tickled to death. It was he who discovered that the white stuff came off, like fine powder. He picked up a white worm from the floor, and drawing it through his hand, showed us that his palm was white. I did the same thing with a leaf.
"It's a deposit of some sort," declared Donald.

"Of what?" asked Biggles.

"Ah, that's what we've got to find out," answered Donald, smelling the stuff. "I think it's metallic," he decided. "Our best plan would be to collect a piece of the rock and take it outside into the light. We shall then be able to examine it more closely."

As he spoke he knocked a chip off the wall and asked me to carry it outside. When I went to pick it up I got the shock of my life. Once, when Biggles was running a transport company, I helped him to carry some gold, and that's pretty heavy stuff... But this stuff... well, I had only to lift a small piece, but it took me all my time to drag it outside.

The first thing we noticed was the colour. The part that had been exposed was still white, but the new face, where it had been chipped off, was red-a rich, glowing crimson. There was no doubt that it was a metal of some sort. Donald started babbling about something called orichalcum...

Ginger then described how the sample became hot and started smoking on exposure to the air. He continued,

Biggles said to me, "What are you looking so scared about?"

I answered, "Nothing." As I spoke, I looked at his face and saw that it was chalky white. "You don't look too happy yourself," I said.

It was old Donald who realized the truth. "Great heavens!" he said, "the stuff is oxidizing on us. We shall all have a coating of metal on us if we stay here."

The party retreated. The abandoned rock sample set fire to the dry moss and in turn set the cave area ablaze, eventually fusing the mountain rockface into a solid mass. Ginger concluded,

Whether the orichalcum inside – if that, in fact, is what it was – had burnt itself out, or whether it remains, nobody knows, and unless an earthquake occurs to expose the cave again, nobody is likely to know. If it's still there, as far as I'm concerned it can stay there. I've no desire to have an armour-plated hide.

EDITOR`S NOTE: Orilachium is the metal described by Plato and others as having only existed on Atlantis. Interestingly it appears that Johns was interested in the legends of Atlantis because he brings up the subject of the mythical metal again in the first two volumes of his much maligned (but rather fun) science fiction series *Kings of Space* and *Return to Mars*.

RICHARD FREEMAN'S NOTE: Animals that have a cthonian lifestyle are often pallid of hue. The lack of sunlight eventually leads to the atrophy of eyes. Without the sense of sight, and in a totally dark world, the need for coloured skin is lost. Some dramatic examples of these eyeless troglodytes include the olm *Proteus anguinus*, a blind, aquatic salamander of southern Europe and the blind cave fish (family *Amblyopsidae*).

Recently whole cave ecosystems, some that have been existing in isolation from the outside world for 5 million years, have been discovered. Here slugs, worms, and shrimps, are preyed on by sightless spiders and water scorpions. These insular ecosystems are the most like Dr Duck's Bornean cave.

Biggles - Charter Pilot: The Silent Death

The flight crews are discussing Dr. Goebells' claim that Nazi Germany has a secret weapon and the consensus is disbelief. Ginger takes issue with this:

"This has been a war of secret weapons. The German magnetic mine was a secret weapon. Our radio-location [radar] *was a secret weapon. Why, even animals and insects know the value of a secret weapon."*

"Here, just a minute, young fellow-me-lad," put in Lord Bertie Lissie. "You aren't by any chance suggesting that jolly old gnats and things sit down and workout new ideas?"

"Well, I'll admit I didn't exactly mean that, but I don't see why not," returned Ginger slowly. "Scores of animals, birds, fish and insects, have what must have been originally a secret weapon. A snake injects poison, the skunk squirts stinking stuff, the armadillo is really a living tank. All of them must at some time have taken other creatures by surprise."

"But all these things are known now; they're no longer secrets," argued Henry. "Animals are no longer capable of developing secret weapons. Only men can do that now."

"I call that a bold statement," declared Ginger. "How do you know what wild creatures are doing? Are you in a position to assert that there is not, in some tropical jungle, even at this moment, a creature that is slowly but surely developing some deadly weapon?"
"No, I'm not," confessed Henry.

"I should jolly well say not," replied Ginger. "Had you done so I would soon have squashed your argument, for, as it happens, there is ample proof that it is not beyond the power of wild things to produce lethal weapons. I know it to my cost."

"What did it cost you?" inquired Bertie.

"It jolly nearly cost me my life." Ginger jerked his thumb in the direction of Algy Lacey. "If you don't believe me, you ask Algy. As a matter of detail, the Silent Death, as the thing was called, put our friend, Dr. Duck, in hospital for two months, and, incidentally, put an end to our biological cruise. For some time it was touch and go with him."

"The Silent Death'?" whispered Henry. "That sounds terrific."

"It was terrific," agreed Ginger,. "Even now I can't think about it without a shudder."

Valiantly suppressing shudders, Ginger proceeds to not only think about it, but to describe what was happening on the West Indies island of Hispaniola – now divided between Haiti and the Dominican Republic...

...In a certain district of Dominica cattle had been found dead in mysterious circumstances. After their death their bodies were found to be entirely drained of blood. This ghastly business always took place at night.

The first obvious answer to this sinister mystery was that the animals had fallen victim to the notorious bloodsucking vampire bats that are common in the West Indies and Central America. But the vampire bat doesn't kill. It takes a meal from the victim, animal or human being, without awakening him. Cattle, goats, and even humans are subject all their lives to attacks by these pests, apparently without ill effect, although too much of it certainly weakens the victims. In short, the thing is so common that it is accepted as a matter of course, as we take wasp stings in this country.

The next point is, the vampire doesn't drain its victim. Obviously, it would be impossible for a creature the size of a vampire bat, which is only about four inches long, to consume the entire contents of an animal the size of a cow. Nor could it be a question of attack by a swarm of vampire bats, because they operate singly. What creature, then, had done the mischief? Nobody knew, but the natives, in their superstitious terror, ascribed the Silent Death to supernatural causes. Time went on, and things got worse instead of better. Moreover, the plague might spread to other islands, for already the outlying farms were affected.

The thing came to a head when first one native, then two or three, died in the same mysterious circumstances as had the cattle. They were also struck down during the hours of darkness. Stark terror descended on the island...

Dr Duck, Biggles, Algy and Ginger fly out to investigate. Sitting out in the open, as bait, is Dr Duck. The others are watching intently. Ginger describes how he saw a few bats flitting around…

...It was nothing to be afraid of. In fact, I was amused. One of the little bats suddenly swooped down and hung on to a dead stalk quite close to my feet. From there he hopped to the ground. Then he began what seemed to be an extraordinary series of antics - still without alarming me. Looking like a little fat mouse, he began to run towards me in a series of sideways rushes, each rush bringing him nearer. At last he was so close that I could see his tiny eyes shining. They glinted red in the moonlight. At that stage my impression was that either the little creature had hurt itself, or else it was looking for worms on the ground. Then, suddenly, without a sound, it took wing and hovered in front of my face, about a yard away. Its wings moved so quickly that I couldn't watch them. It was rather like a humming bird, except that this little creature swung slightly from side to side. Even then all I thought was what an extraordinary thing to do. I wondered what it was after. Not for an instant did I suppose that it was after me

I find it difficult to describe what happened next. I've never been a believer in hypnotism, but the effect of this swinging in front of my eyes must have been hypnotic. I just sat and stared; and as I stared the use went out of my limbs. It was like having gas at the dentist's. I could feel myself going off. The dickens of it was, I could do nothing about it. I suppose I was too far gone. Then in a vague sort of way I was aware of the bat coming closer. Then the fluttering ceased, and I knew it had settled on me. There was a tiny prick in my throat that might have been made by the sharpest needle imaginable. And that's all I remember...

The next thing I knew I was lying on my back)n the camp bed in our house, with the others round me. Donald, in his shirtsleeves, was holding a bottle near my nose. I heard him say, "It's all right, he's coming round..."

Ginger recounted Dr. Duck's explanation of the matter:

"We know now that the thing is a bat, evidently a new species. First it mesmerizes its victim, in much the same way that a hypnotist mesmerizes a subject by passing his hands in front of a person's face. Then it settles, and reduces the victim to complete unconsciousness by means of a fluid injected through a tiny tube, in the manner of a hypodermic needle. It then proceeds to satisfy its hunger, in which nasty business it is joined by hundreds of others."

Biggles, Ginger said, wanted to find and destroy the colony of bats before they spread any further. Stumbling across their 'headquarters' by accident, the bats attacked the party and they ran for it. The natives of the island, observing this, waited until things had settled down and then blocked the cave with brushwood and set fire to it.

It is notable, incidentally, that W. E. Johns was interested in having his heroes make potentially world-changing discoveries but would then, for the sake of future contemporary consistency, engineer some sort of wipe-out of the evidence, so that nothing remained of the discovery but anecdotal evidence...

<u>RICHARD FREEMAN'S NOTE</u>: Here the author seems to have predicted the chupacabra about 50 years before its advent! Of course no bats actually hypnotise their prey. Many animals from snakes to weasels are believed to mesmerise prey animals. This is in fact the prey freezing in an attempt to not be seen or to avoid a strike. The vampire bats (*Desmodus spp*) take only tiny amounts of blood. They open the skin with sharp incisors and lap the trickle of blood. Their saliva contains an anti coagulant that keeps the blood form clotting for a short period. To totally drain a prey animal would be detrimental for the bat's future feeding.

The true vampire bat is found only in the neo-tropics but it may have an old world counterpart in Africa. Dr Karl Shuker has noted that in 1936 archaeologist Byron de Porok was told of wing monsters called death birds whilst travelling in southern Ethiopia. These were supposed to have killed several goat herders who had slept in the vicinity of Devil's Cave near Lekempti. On entering the cave Porok disturbed a swarm of bats that the locals told him were the death birds. They left their victims covered in puncture marks, with bloodstained clothes, and drained of blood. Porok himself saw one man close to death and looking like a living skeleton. This seems an odd strategy for blood drinking bats to evolve. Perhaps some kind of disease was the cause of death and the bats wrongly blamed. Maybe an Ebola type virus was resident in the cave system like it was in Uganda. Only further expeditions will tell

Radium and centipedes: Biggles Hits the Trail

Back in the 1920s the intelligentsia was becoming excited by the strange properties of a rare material called radium. This was an elemental material first discovered in 1898 by Marie Curie. After the First World War scientists managed to isolate a few grams and begin to investigate its properties. If held in the hand, it gave off spontaneous heat, and, almost incidentally, caused curious burn damage to body tissue on longer exposure. It's potential for cancer therapy and even as the source of a possible death ray spawned some lurid press stories in the 1920s and 30s.

In our post-Hiroshima, post Cold War society, it can take quite an effort of imagination to understand the popular enthusiasm that greeted the discovery of a radioactive material. Back then though, it was a different matter, and many journalists and authors had their imaginations fired up by the news of this miracle metal.

While Agatha Christie busily challenged Hercules Poirot to deal with "*The Big Four*" (1923), Johns wrote "*Biggles Hits the Trail*". In retrospect, both radium-inspired books are the respective authors' most overtly science fiction excursions - but one has to remember the climate of opinion at the time: the pundits were enthusing wildly about radium and, really, the books are simply both products of their age; a brief age when, to many, it seemed that science and society was on the brink of major beneficial change.

It was Johns, though, with his occasional *penchant* for science fiction stories (featuring characters other than Biggles) that extrapolated radium's then-hyped abilities the further and, in so doing, gave Biggles one of his nastiest moments.

Biggles, Algy and Ginger fly to the Mountain of Light in Tibet, home to a mysterious race, the Chungs. Accompanying them is Dickpa, Biggles' uncle, who believes that the mountain contains radium deposits. He and a medical colleague, Lord Maltenham, hope to collect some samples for research. On the final stage of their journey, approaching the mountain on foot, they make a find.

The white object was a dead centipede of huge proportions. Biggles thought it was the most repulsive object he had ever seen. It was about the size of a sausage, bloated in a disgusting way; its toad-like skin of fish-belly white glowed with a metallic lustre. Black beady eyes glittered banefully, even in death; and from a cruel shark-like mouth protruded needle-like, in-curved teeth.

The next day, and, centipede forgotten, the party are making their way on foot to the upper parts of the mountain. All except Ginger are finding the going increasingly difficult; eventually the other four are brought to a halt through creeping paralysis of their limbs. The ground and rocks appear mysteriously electrified. It's realised that Ginger is the only member of the party wearing rubber-soled shoes and he runs back to the aeroplane to collect rubber waterproofing sheets from the tent. While Biggles and the others are forced to await Ginger's return, Biggles' attention is caught by a movement.

On the far side of the arena yawned the black mouth of a cave, rather larger than any of those they had seen in the gorge. From it, something appeared to be flowing, a dull white stream above which hung a pale blue mist. The movement was not regular, but undulating, like a shallow stream of milk flowing over a bed of large stones. It was coming in their direction, accompanied by a curious faint crackling sound.

"What is it?" breathed Algy.

"Can't you see," muttered Biggles desperately.

"It's centipedes," cried Algy, in a strangled voice, suddenly understanding. With a colossal and instinctive effort to escape, he forced himself round to face the path, but two dragging steps was sufficient to show that flight was out of the question. "What can we do?" he cried, in a voice tense with horror.

They could see the reptiles now distinctly, as they surged over the ground in a sinuous rippling movement that began at the head and ran down to the tail. Their mouths were wide open, like hunting wolves...

"It's the first time in my life I've had to admit it, but I don't know," replied Biggles. "I couldn't make ten yards, not even with those behind me. If Ginger's another three minutes I'm afraid he'll be too late," he added in a voice that was strangely calm.

Happily for our heroes, Ginger returns in time to allow everyone to clad their feet in sheet rubber and, thus insulated, they put a safe distance between themselves and the creeping horror before adjusting their hastily-improvised footwear.

Biggles finished first and looked back up the gorge. What he saw seemed to surprise him, for he took a pace or two nearer and continued staring. "Well I'm dashed," he muttered. "What do you make of that?"

The centipedes had stopped moving, and on closer inspection were all found to be dead. Indeed, the party were able to continue their trek, squelching their way through the centipedes.

Eventually Biggles meets a Mountain of Light engineer, a stranded Westerner, who explains a few things about these mysterious centipedes.

"You said something about these devilish things being electrically controlled," suggested Biggles.

"That's right. They've got funny ideas, the Chungs - very peculiar ones. They don't think as we do. They wanted some sort of defence for keeping inquisitive people out, so they worked their plans on their own lines. They don't understand guns, or anything like that. The centipedes is one idea. Hundreds of years ago the scientists got to work - clever devils they are, too - and in the end, by hybridizing snakes and scorpions and all sorts of horrors, they produced a new sort of centipede. Poisonous, they were. But one day they all got loose and pretty near killed the whole colony before they were rounded up again. The Chungs saw that a weapon that could turn around and bite them wasn't much use, so they started off afresh and produced a new lot. They're not poisonous but they just eat anybody alive by sheer numbers. All the same, odd ones were always escaping and the people got fed up with being bitten, so they started off on a new scheme. They produced a breed with funny sort of pads on their feet that could only move when the ground was electrified. Wherever they were, and whatever they were doing, they could paralyse the lot by a single switch."

Biggles commented that since he didn't understand the centipedes and the Chungs didn't understand guns and explosives, that more or less evened things up - and the expedition then proceeded on that basis and, after many tribulations (including encountering death rays) eventually collected some radium-bearing rock samples, destroyed the Chungs' lair, and returned home.

RICHARD FREEMAN'S COMMENT: W.E. Johns must have been smoking something when he wrote this! He also shows that he is no zoologist, calling centipedes reptiles! In real life giant centipedes do exist.

The tropical and sub-tropical family *Scolopedridae* can grow to over a foot long and inject potent venom through the modified first pair of legs (not the mandibles). Though rarely fatal the venom is highly painful. There are stories (probably apocryphal) of victims hacking off their own arms due to the pain.

During the early Carboniferous period, some 360 million years ago giant millipedes existed. Fossils and fossil trackways seem to indicate an animal of 6 feet in length and as broad as a man's leg. These huge herbivores would have been among the largest land animals of their time. We do not know if centipedes ever reach this size but it is interesting to speculate that the giant millipedes had an equally large predator. A 6 foot centipede would be a formidable killer and one can readily imagine them trundling through the forests of giant horsetails and ferns with their furtive feelers twitching.

Due to competition from vertebrates land-based arthropods of this size no longer exist.

The fact that they once *did* however seems to show that there is no biomechanical reason to stop them reaching these dimensions. With the science of genetics advancing apace perhaps we will see truly giant centipedes in tomorrow's laboratories. A horde of such Lovecraftian monsters would make an awesome bio-weapon, highly venomous and psychologically disturbing.

Give us a toke, Captain!

Biggles: Charter Pilot – The Enchanted Island

Another Donald Duck adventure narrated by Ginger in an off-peak moment during the Second World War was sparked by comments from a colleague, Henry Harcourt, about the wartime press. "You can believe nothing you read and nothing you hear," he complained. Ginger said that, on the contrary, he could believe anything, and described his adventure on a volcanic island in the Atlantic by way of illustration. The island had just emerged from the sea...

"There was more to see than I expected. The place was littered with shells. Oysters - you never saw such monsters in your life. They were the size of drums. The mussels were as big as bath-tubs - one would have made a meal for a squadron of men. They were still alive, too, and I took care to keep clear of them. Of course, this should have warned me that everything was likely to be on the same scale, but it's easy to think of these things afterwards.

Ginger described how he was gazing into a rockpool when he heard a scuffling sound behind him and looked around.

"Coming towards me was the father of all crabs. You couldn't imagine such a brute. The shell was a good five feet across; and the two big claws - well, they were big enough to tear a man in halves if they got hold of him. It made a noise like a grandfather clock ticking, only louder. Its eyes, which were black, and stuck out on things like rods, were fixed on me. I started to back away, but as soon as I moved it came on with a rush. I went off with a rush, too, yelling for Biggles.

"Fortunately, Biggles and Algy weren't far away. They heard me yell, and came at the double to see what was going on. As long as I live I shan't forget Algy's face when he saw that crab.

Biggles shouted to him to fetch the emergency axe from the machine, and presently he came back with it and the rifle which we always carried. Biggles took the axe.

He was only just in time, too, for the brute was close behind me. When it saw that there were two of us it got into a flat spin, as if it couldn't make up its mind which one to go for.

It decided on Biggles. I heaved a rock at it, but as it happened Biggles didn't need any help. He jumped aside as the crab made a grab at him, and at the first -slash with the axe severed one of its long claws.

It was horrible to see the way the claw went on opening and shutting after it had been cut off. Then, running in on the clawless side, Biggles sank the axe in the middle of the brute's shell. The crab spat at him, and you never smelt such a foul stink in your life. Anyway, it had had enough, and blundered off into a pond taking the axe with it. I don't mind telling you that when we looked at each other we were all pretty green about the gills."

EDITOR'S NOTE: The incident with the land crabs is very reminiscent of an event described in another children's book from about the same time. Arthur Ransome is not known for his descriptions of horrific wildlife. Indeed the nearest to zoology that most of his excellent children's books comes is in accounts of bird watching on the Norfolk Broads. However, in the third of his *'Swallows and Amazons*` series; *"Peter Duck"* he describes a series of encounters with crabs on a Carribean island that are very reminiscent of those described in this story by Captain W.E.Johns.

Realising Dr. Duck might also be in danger, they set off to look for him, and find him being pursued by a strange object:

"It was a great grey slobbering mass the size of a barrage balloon - or it looked that size to me. The colour was elephant grey, and in fact, it might have been an enormous elephant without any legs. It seemed to roll and bounce over the ground. Around it were what appeared to be a mass of snakes. They were its tentacles, for the thing was a great octopus - or rather, according to the Doctor, a decapod, which has ten arms instead of eight. The two front ones were the longest. I've seen an octopus with tentacles twenty feet long, but these must have been nearer fifty, and they were covered with suckers, like dirty plates.

"Well, Biggles grabbed the rifle from Algy, and dropping on one knee, opened fire. Knowing how Biggles can shoot there was no doubt that he hit it, but he might as well have aimed at a tank for all the effect the bullets had. The monster was overtaking Donald, and it seemed to me that nothing could save him. Biggles dropped the rifle, tore to the machine, and came back with a spare can of petrol, unscrewing the cap as he came... Biggles ran straight by us towards the Doctor. As soon as he reached him he swung the petrol can round so that it splashed spirit all over the place. Then he dropped a match on it and bolted. It was about time, for the two leading tentacles were within a few feet of him. The petrol went up with a whoosh, singeing Biggles's eyebrows. He didn't wait to see what happened to the beast, but turned and bolted."

In fact, the entity retreated, "screaming like a frightened horse". However, the exploration party's problems were not over yet, as the island then started to sink, making a hasty and fraught takeoff necessary.

Ginger turned to Henry. "Now perhaps you understand what I mean when I say that it doesn't do to disbelieve a thing because you haven't seen it yourself. Wait till I tell you about..."

Biggles broke in. "I think that will do for one evening," he said, looking at the clock. "It's time we went to bed. The squadron's on early patrol in the morning, don't forget."

RICHARD FREEMAN'S COMMENT:

The largest land dwelling crab, and indeed the largest terrestrial arthropod is the Christmas Island race of the giant land crab also called the robber or coconut crab *(Birgus latro)*. These immense crustaceans grow to the size of a small dog and are the supreme predators on Christmas Island. They are omnivores feeding on carrion, plant matter (including coconuts that they crack with their powerful pincers) Natives tell (unproven) stories of them killing humans who sleep in the open, unprotected on the ground. The tales tell of the crabs cracking human skulls and eating the brain matter.

An even larger crab lives in the sea. The Japanese spider crab *Macrocheria kaepferi* of the deep waters off southeastern Japan has a leg span of over 12 feet. Its limbs, as the name suggests are elongate. The body can be as far across as a car tyre. On land the crab`s gangly legs and great weight render it helpless.

W.E. Johns was not exaggerating when he described the titan octopus that pursued Dr Duck. Such tentacled abominations do, in reality crawl across the seabed. On the 30th November 1896 a massive carcass ran aground on a beach near St Augustine, Florida. A local surgeon and naturalist Dr DeWitt Webb examined the cadaver finding it to be 20 feet long and 4 feet high. It consisted of a huge pear shaped, sack like body with 8 stumps of differing lengths attached to it. It also had wing life fins or flanges. The whole mass weighed an estimated 7 tons. The skin was thick and very difficult to penetrate. Realising this was of scientific interest the Dr had a team of horses drag the body further up the beach.

Webb wrote to Professor Addison Emery Verrill and sent samples of the monster's tissue. Verrill declared the creature to be a gigantic octopus. In life, he reckoned it would have had a tentacle span of a terrifying 200 feet! With the added weight of the tentacles the brute would have tipped the scales at 20 tons.

Later, perhaps with his mind recoiling from the idea of such a horror Verrill did a scientific U-turn and declared that the Florida carcass must be the rotting remains of a sperm whale. And there the mater stood until Webb`s long forgotten samples turned up in the collection of the Smithsonian Institute in the early 1970s. Unearthed by Forrest Glenn Wood, a marine biologist, after a 13-year search. They were subsequently examined by Joseph F.Gennaro Jr, a biologist at the University of Florida. Gennaro examined the samples under polarised light to compare the connective tissues to control samples. The results showed that the tissue was indeed from an octopus.

Fishermen in the Bahamas know of such giants. They call them the Lusca, or hairy hands. The latter name refers to a fringe of hair like filaments that run down the animal's tentacles. They are said to inhabit deep caverns on the seabed known as the blue deeps. The Lusca is said to be highly dangerous.

The filaments on the tentacles and the wing like fins point to the Lusca being a giant cirrate octopus. The comparison with H.P.Lovecraft's horrific Great Cthullu is irresistible. A "winged" tentacled atrocity inhabiting the strange geometry of the blue deeps! Art imitating life or visa-versa?

Biggles: Charter Pilot – The Patagonian Giants

On the few occasions when strange discoveries made, they were generally cryptozoological animals. However, one adventure was based on the legends of the Giants of Patagonia (the southernmost part of South America). Ginger commented on the background to the legends.

Admittedly, in those days some pretty tall stories were told, but there seemed to be no reason why mariners should invent a race of giants if they didn't really exist. It wasn't as though only one ship's company saw them.

Scores of people saw these giants, and in a good many official log-books we find descriptions of them. It is not surprising, therefore, that these giants became an accepted fact. But later, when steamboats appeared, and scientists Went out to look for the giants, they couldn't be found. What happened to them nobody knows.

This is a mystery that has puzzled scientists ever since. It puzzled Dr. Duck. That these giants existed we need not doubt. Master Francis Pretty, who went round Patagonia with three ships under General Thomas Candish, and afterwards wrote an account of the voyage, says that they caught one of these giants, and, by measurement, found that the size of his feet were eighteen inches long. That, in proportion, would make the man about nine feet tall. As a matter of detail I saw some who must have been taller than that.

Mind you, I knew nothing about this until Dr. Duck came in one day and gave us the benefit of

all the information that he had collected on the subject. His idea was, as you may suppose, that we should go and have a look for these lost giants. And that was really the object of our voyage, although it turned out somewhat differently from what we expected. At any rate, we went, and I may say that it was no picnic. I've never been so cold and miserable in my life as I was in Patagonia. I don't mind clean, hard frost, but the icy winds nearly got me down.

Ginger described how the exploration party had a forced-landing due to engine trouble and the next thing they knew, they were being approached by 9-ft (3 m) tall giants, who, happily for the visitors, were not aggressive and, furthermore, spoke rudimentary English. Invited back to their caves for soup, the party found that the concoction enhanced strength and their bodies soon began to enlarge. Meanwhile, Dr Duck was excited to find the tribe had some documentary archives.

Donald said they were pages out of an old sea log-book. The writing was early English, with a lot of Norman-French and Latin mixed up with it. There was a crude chart, lettered in Spanish, which bore the date 1381.

"That means," Donald pointed out, "that these people were the real discoverers of America - certainly the first colonists. You realise that if this date is correct, and there is no reason to doubt it, the ancestors of these people must have landed here a hundred years before Columbus, who didn't cross the Atlantic until 1492. We know from William of Worcester's Annals of England, dated 1324, that the existence of the New World was known long before Columbus's time. He says that two ships went out from Bristol to find the island of Brazil - and that was not an isolated venture. There seems to be no doubt whatever that in 1381 an English ship got across. Probably it was wrecked on the coast. At any rate, the sailors couldn't get back, and they must have settled here, or, in the first place, near the coast, and intermarried with the natives. When the first Elizabethan mariners reported giants here they little knew that they were talking about their own countrymen. They were too scared to land and make inquiries."

The giants had no idea of their origin. As far as they knew, they had always dwelt there. Their diet was pathetically frugal, for the inhospitable country offered practically nothing in the way of food. They had two main dishes, moss and a kind of fresh-water mussel which they found in the lake. These were stewed together, and it was the mixture that gave them strength. So, by the curious irony of fate, instead of starving to death as you might suppose, they had accidentally struck a diet which not only offered the vitamins necessary to support life, but increased their stature and their strength to a tremendous degree. They themselves were unaware of this because, as far back as they could remember, they had never made contact with other people.

Having fixed their aircraft, the exploration party took their leave of the gentle giants of Patagonia. Ginger concluded by describing how a return trip the following year discovered that a landslide had buried the giants' caves and filled the entire valley.

Ginger yawned and looked up. "Well, that's all. I'm going for a stroll to get a spot of fresh air. Anyone coming?"

RICHARD FREEMAN'S COMMENT: Antonio Pigafetta, who wrote a journal of Ferdinand Magellan`s round-the-world voyage stated that Magellan's crew had met giants in Patagonia. "This man was so tall our heads scarcely came up to his waist and his voice was like that of a bull," he wrote. Another explorer, Jacob Lamaire wrote that on December 17th 1615 he had found giant's graves. At Port Desire in the Straits of Magellan, several stone covered graves were discovered. The skeletons within , he said, measured 10-11 feet in height. This was a generation after Magellan's expedition.

Tierra del Fuego, at South America`s tip seems to have been a giants strong hold. On ship gave an account of landing there and trading trinkets with the giants for fresh fruit and vegetables. They averaged on 8 feet tall, were clothed in sheepskin and had black hair. The crew were amazed to see they had no huts or dwellings to protect them from the harsh climate. Indeed there seemed to be no trees.

Magellan apparently tried to capture a giant on a previous voyage in 1519. Nine men were insufficient to hold the giant, who escaped. Sir Francis Drake had better luck and managed to capture some youngsters, one of whom he taught English. The child told him that there were four tribes on Tierra del Fuego. The tribes were constantly at war. What became of these unfortunate children is unknown.

In October 1766 French Naval captain Alexander Guyot wrote of meeting the giants. Some seemed willing to return to Europe with the crew. However once on board they burst into tears. They were returned to shore but the misunderstanding caused a fight in which several giants were killed.

Chenard de la Girandais captain of the *Star Pink* reached the Straits of Magellan in May 1799. He wrote of seeing 700-800 savages of great stature. Some were 6 feet tall. It seems these cannot be the same people as his predecessors had seen.

Darwin was the last European to report on the giants. In his record of the *Beagle's* voyage he records meeting friendly natives at Cape Gregory. Their average height was about 6 feet. Darwin believed that the heights previously reported were exaggerations due to the native's long hair and guanaco hide mantles.

This seems to great a stretch of the imagination, hair and clothes cannot double a man's height! Perhaps the true giants had died out. This could have been due to diseases transmitted by European mariners coupled with their own in-fighting. An archaeological expedition to Tierra del Fuego will hopefully this riddle once and for all.

References:

Biggles Charter Pilot ©1943. 1965 Armada Paperbacks (Hodder & Stoughton) edn.
Biggles Hits the Trail ©1935. Brockhampton Press, Leicester UK.
Biggles and the Cruise of the Condor ©1933. Red Fox paperback (Random House UK) 1994.

Cassowaries

THE MEANEST, COOLEST BIRDS ALIVE (PERHAPS)
Text & Illustrations © Darren Naish

1. What cassowaries are

Cassowaries are large, black, emu-like flightless birds with distinctive horn-covered head crests and brightly coloured, fleshy wattles that hang from their necks. Their wing feathers (remiges) are elongate, stiff quills and they sport an enlarged, dagger-shaped claw on their second toe. The word cassowary is apparently derived from the Papuan words for horned ('kasu') and head ('weri'). Known only from the rainforests of northern Australia, New Guinea and the surrounding islands, cassowaries are famous for their reputed evil temperament. They are strong swimmers and are quite capable of crossing large rivers. This might partly explain their occurrence on some of the smaller islands in the Australasian region. However, the main reason for their occurrence on islands such as Ceram and New Britain is almost certainly introduction by native peoples.

While both sexes in all species of cassowary share the same kind of wattle adornment, female cassowaries are larger and more brightly coloured than males (the latter is arguable), and also have bigger casques and larger second toe claws. Emu and kiwi females are also larger than males, though in ostriches and rheas the reverse is true. Male cassowaries look after the eggs and babies, while females are thought by some to possess a penis-like structure (see below). The birds therefore appear ambiguous in their sexuality and are regarded as androgynous or hermaphroditic by the Australasian peoples who have observed them. Cassowaries have since become important characters in the mythology and iconography of the people of New Guinea and elsewhere. They use these birds as important commodities, providing feathers, meat and trade. In this article I review cassowary natural history, ethology, evolution, systematics and more – in effect, everything you wanted to know about cassowaries but were afraid to ask.

2. Cassowary ecology and reproduction

Cassowaries are predominantly rainforest birds and they rely on the year-round availability of large numbers of fruiting plants. The cassowary breeding season (July-October) coincides with the peak fruiting season. Consequently, cassowary populations are highest in areas that support the greatest diversity of fruiting plants (thus ensuring year-round fruit availability). Virgin forest is not essential for cassowaries and they can do well in managed forests as long as plant diversity remains high. Furthermore, ideal cassowary habitats may be patchworks of open woodland, sclerophyll forest and rainforest, rather than rainforest alone. Cassowaries have also been reported to enter mangroves.

Like most other ratites, cassowaries forage for low-growing plants, pick up fruits from the ground, and also eat small animals. The bulk of their diet consists of berries and other fruits, particularly of Lauraceae species, and they have been recorded consuming the fruits of 75 different species. Cassowaries also supplement this diet with fungi, insects, land snails, and reputedly with leaves (Burton 1984), small vertebrates and some carrion. Among the vertebrate material cassowaries have been recorded eating are skinks, birds and rats. They swallow fruits whole, even very large ones.

Fruit seeds are not broken down as they pass through the cassowary's digestive system and some seeds, and indeed even some entire fruits, may be excreted in an undamaged state. Because these seeds are still able to germinate, the plants that produce them may benefit from the cassowary's behaviour, their seeds being deposited perhaps several kilometres from their original location.

It has been suggested that several Australasian rainforest plants rely on the digestive effects of the cassowary gut to trigger germination and as many as 21 such cassowary-dependent species have been identified (Stocker and Irvine 1983). Cassowaries therefore have a key role to play in the maintenance of the ecosystems that they inhabit. Evidence that cassowaries are metabolically dependent upon the protein-rich laurel fruits on which they feed comes from the fact that cassowaries breed less successfully in years when laurels fail to fruit.

> NB – the well-known suggestion of Temple (1977) that the Mauritian Tambalacoque tree (*Sideroxylon grandiflorum* [formerly *Calvaria major*]) relied on a relationship of obligate mutualism with the dodo (*Raphus cucullatus*) (e.g. Durrell 1977, Shuker 1997) has been effectively counter-argued by Owadally (1979) and Witmer and Cheke (1991) and is not regarded as likely by those who work on the palaeocology of Mauritius.

Cassowaries are solitary birds that maintain territories of 1-5 km^2 and, outside of the breeding season, probably most interactions between cassowaries are aggressive. When males encounter one another, they stretch, fluff out their feathers, and make rumbling calls until one of them backs down. Fights are initiated by a bird raising its feathers, bending its neck beneath its body and roaring. Males fight by kicking and leaping (Fig. 1), though their bouts are normally

Fig. 1. A Double-wattled cassowary (*C. casuarius*) leaping and kicking, the elongate claw on the second toe being deployed as a weapon. At right, medial view of a cassowary left foot. Note that there is no hallux (first digit). Main illustration based on a painting by Guy Tudor.

brief and the birds tend not to be injured. Females are dominant to males and a female can usually intimidate a male by simply staring, by rumbling, or by stretching.

Male cassowaries typically flee from females following such encounters. During the winter breeding season, each territory is occupied by a pair. Males court females with a low, booming call and also dance around the female prior to copulation.

The nest is a shallow scrape, sometimes lined with a few leaves, and the female leaves the nest site immediately after laying. Females may be polyandrous (mating with several males), mating with two or three males in a breeding season. The 3-8 eggs (usually 4) are incubated for 47-54 days by the male. He then guards the chicks for about 9 months and normally chases them away after this time (Crome 1975). Juveniles have head-to-tail stripes, become dull grey or brownish after about 3 months, and grow their glossy black adult plumage after their first year. Male cassowaries that have yet to grow their full adult plumage may attempt to court females and are sometimes successful.

2.1. Sexing cassowaries

Native myths and legends about the androgynous nature of cassowaries are supposed to be supported by the presence in both sexes of a penis-like organ (Bagemihl 1999). Whether this is true or not depends on which source you consult. Some authors argue that females categorically lack any penis-like organ and that cassowaries can be reliably sexed by feeling for a phallus within the proctodeum of the cloaca. This rests in a pocket on the ventral surface of the proctodeum. However, according to others, all cassowary chicks possess what appears to be a flaccid phallus. Among adults, this organ can be seen when they defecate and is notably smaller in females than in males. Having never carefully observed or probed the cloaca of a cassowary I cannot resolve this apparent conflict in opinions. A similar debate, with similar conflicting opinions, exists in the literature on the sexual organs of male tuataras (*Sphenodon*) – do they or do they not have an intromittent organ?

The truth appears to be that, while *Sphenodon* lacks a penis or hemipenis, it does possess cloacal pouches that are everted during copulation (Cree and Daugherty 1990).

3. The living cassowary species

Cassowaries are represented today by three species, all of which are closely related and rather similar, though adapted for different rainforest habitats. Cassowaries have a very complicated taxonomic history and numerous species and subspecies have been named – for an extensive review see Rothschild (1900). Given that individuals from some islands have apparently been transported to others, and that hybridization has occurred between many of the so-called subspecies, it is a subject too complex to cover here. Rothschild recognised eight species that he grouped into three assemblages. There were the typical cassowaries (*Casuarius bicarunculatus* and *C. casuarius* – the latter with seven subspecies), the one-wattled cassowaries (*C. philipi* and *C. unappendiculatus* – the latter with four subspecies) and the mooruks (*C. papuanus* with two subspecies, *C. picticollis* with two subspecies, *C. bennetti* with two subspecies and *C. loriae*). Subsequent taxonomic work has meant that Rothschild's three 'assemblages'

Fig. 2. Double-wattled cassowary (*C. casuarius*). This individual (drawn from a photo) sports a notably tall, rounded crest.

Fig. 3. One-wattled cassowary (*C. unappendiculatus*). The crest in this species tends to be lower than that of *C. casuarius*.

Fig. 4. Dwarf or Bennett's cassowary (*C. bennetti*). As with the other two species, wattle and skin colour is highly variable in this species.

Of these three, the largest and best known is the common, double-wattled, Southern, Australian or Ceram cassowary (*Casuarius casuarius*) (Fig. 2).

Several other species named over the years, including *C. galeatus*, *C. orientalis* and *C. javanensis*, are presently regarded as junior synonyms of *C. casuarius*. Growing to 2 m in height and 85 kg, this is the most widely distributed species and the only one found on the Australian mainland (in the rainforests of north-eastern Queensland) and was probably originally a native of mid-level rainforests. In Papua New Guinea it lives along the northern coast, on the lowlands of the south side, on the Vogelkop and on the west side of Geelvink Bay. It is also found on Ceram and the Aru islands, probably by way of human introduction (White 1975). The endemic Queensland race of *C. casuarius*, *C. c. johnsonii* (or *johnsoni*), is currently divided into two isolated populations. A third, presently unverified, population may persist at the north of Cape York Peninsula while a fourth, from the extreme north of Cape York Peninsula, became extinct in the 1980s.

The single-wattled or one-wattled cassowary (*C. unappendiculatus*) (Fig. 3), named by Blyth in 1860 for a captive specimen examined in Calcutta (see below), is a probable low-level rainforest denizen. *C. unappendiculatus* takes the place of *C. casuarius* in northern New Guinea from the western Vogelkop to Astrolabe Bay. It also occurs on the islands of Salawati (or Salwatti or Salwatty) and Japen (or Yapen). *C. unappendiculatus* is smaller than *C. casuarius* and is confined to New Guinea. Reaching 1.6 m, its single wattle is small (approx. 3 cm long).

The dwarf or Bennett's cassowary (*C. bennetti*) (Fig. 4), also known as the Mooruk or Moruk (both names are apparently corruptions of the New Britain word 'Moorup'), is the smallest extant cassowary (reaching 1 m) and is variously reputed to be the most aggressive or the most docile, depending on which author you believe. Before its range was disrupted by human interference, it may have been a highland specialist and might even have frequented montane grasslands (Davies 1991). It is found on New Guinea (from the Vogelkop to the Kratke mountains), New Britain (where it has probably been introduced – White 1976) and Japen. *C. bennetti* does not have the tall, laterally compressed casque of the other two species, but has a low mound-like crest instead.

3.1. Some 'forgotten' cassowaries

An assortment of poorly known cassowary species and subspecies has been described in the past but since forgotten as they have been subsumed into synonymy with the three currently recognised species. While the vast majority of these have certainly represented individual variants, growth stages or hybrids of the currently recognised forms, some of them are still of great interest. Others sound extremely curious from the historical documents that describe them and leave one wondering whether they are worthy of further investigation.

As with most animals, melanistic cassowaries occur and all-black individuals of *C. unappendiculatus* have been reported from Jobi Island. These were described in 1893 as the new species *C. laglaizei*, later demoted to a 'variety' of *C. unappendiculatus*. Another species regarded today as synonymous with *C. unappendiculatus*, Sclater's cassowary (*C. philipi*), was named by Rothschild in 1898 for a captive specimen kept in the Zoological Gardens at London. Shipped from Calcutta and named in honour of Philip Sclater (who is also commemorated in the name of a *C. casuarius* subspecies), it was probably captured in New Guinea and is worthy of note because of the extraordinary morphology Rothschild described for it. In fact, to Rothschild, Sclater's cassowary was the most distinctive of all cassowaries.

On naming the species in 1898, Rothschild thought that, despite its brown feathers, it was fully-grown and therefore unlike other cassowaries in colour. Over the years however, its feathers turned as black as those of any other cassowary. However, other of its features remained highly unusual. Not only were its feathers structurally more like those of an emu than of a cassowary, the feathers from its rump and tail region were extraordinarily long – so long that they dragged on the ground. Its casque was described as intermediate between that of *C. unappendiculatus* and *C. bennetti*, being compressed rostrally but mound-like caudally. Its call reportedly 'resembled a deep roar' and was unlike that of other cassowaries. Most remarkably, however, it had notably stout, short legs and, though it was large bird (Rothschild described it as 'a giant'), it was lower to the ground than any other of the large cassowaries, being equal in height to the small Bennett's cassowary. Rothschild even likened Sclater's cassowary to *Pachyornis*, the stout-legged moa, a moa famous for its large size but thickset, short-legged frame. Fascinating as this animal sounds, it is now regarded as an individual of *C. unappendiculatus*. Despite Rothschild's confidence about the distinctive nature of Sclater's cassowary, it was apparently still a juvenile or subadult during the time that Rothschild was describing it. Its bizarre feathers and unusual proportions were purportedly due to individual variation and perhaps its lifestyle in captivity. Nevertheless it sounds like a remarkable bird.

In his writings on cassowaries, Rothschild's greatest mistake was perhaps to recognise distinct species whenever he encountered a cassowary, which had a particularly bold colour pattern on its head and neck. This propensity to recognize multiple species based on small differences was even commented on by his sister-in-law and his employees during his lifetime (Rothschild 1983), and should not be regarded as a criticism unique to this enlightened age. Among these distinctively pigmented forms were the painted-necked cassowaries, the first of which, *C. picticollis*, was named by Philip Sclater in 1874 for a New Guinea specimen kept at London. The descriptions indicate that *C. picticollis* was identical to *C. bennetti*, its distinguishing features being its red throat and bright blue nape and fore-neck. Loria's cassowary (*C. loriae*) was named by Rothschild in 1898 for a short-beaked specimen from the highlands of New Guinea previously referred to *C. picticollis*. This apparently had vivid scarlet on part of its neck and, according to Rothschild, thus deserved to be separated from *C. picticollis*. This species too has not proved distinct from *C. bennetti*, though its reportedly short beak is actually at odds with the rather long and curved beak of *C. bennetti*.

4. Cassowary characteristics and morphology

Cassowaries have black, shiny plumage, which hangs loosely, like long fur (much as it does [or did] in kiwis, moa and emus). Unlike those of emus, the feathers are not double-shafted. It has been suggested that the loose cassowary plumage helps in withstanding abrasion from dense vegetation, and the bony casque on the head is alleged to do so too. The casque as observed in the live bird is a large horn sheath that grows from a smaller bony crest on the top of the skull. The actual bone crest is not solid but hollow, light and supported internally by a honeycomb of supporting bone struts. The bony crest is actually fragile and easily broken. The casque tends to lean over to the right hand side (Fig. 5). This may be the result of use: cassowaries in captivity have been observed using their crests to move soil and leaf debris while foraging and it seems possible that this is an important function of the crest in the wild.

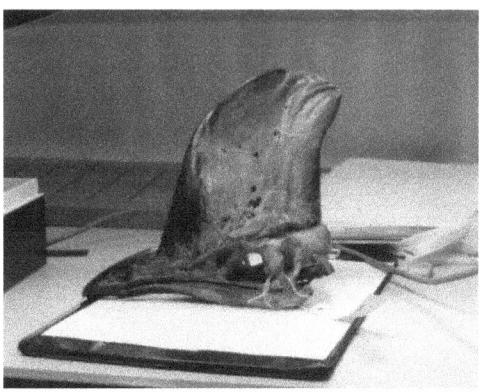

Fig. 5.a   Left lateral view of *C. casuarius* skull with rhamphothecae and casque still attached.

Fig. 5.b   Same specimen of *C. casuarius* skull as pictured above in rostral view showing casque leaning to right hand side.

Besides the casque, which varies from tall and laterally compressed in *C. casuarius* to low and mound-like in *C. bennetti*, cassowaries are also adorned with vivid red, purple, blue, green or yellow areas of naked skin covering the throat, back of the neck and head. On the neck and throat especially, this naked skin is wrinkled and, on the throat, supports red or blue hanging wattles (except in *C. bennetti*). The cassowary bill is gently decurved and robust. The margins of the rostral end of the lower jaw have serrated margins (Fig. 6).

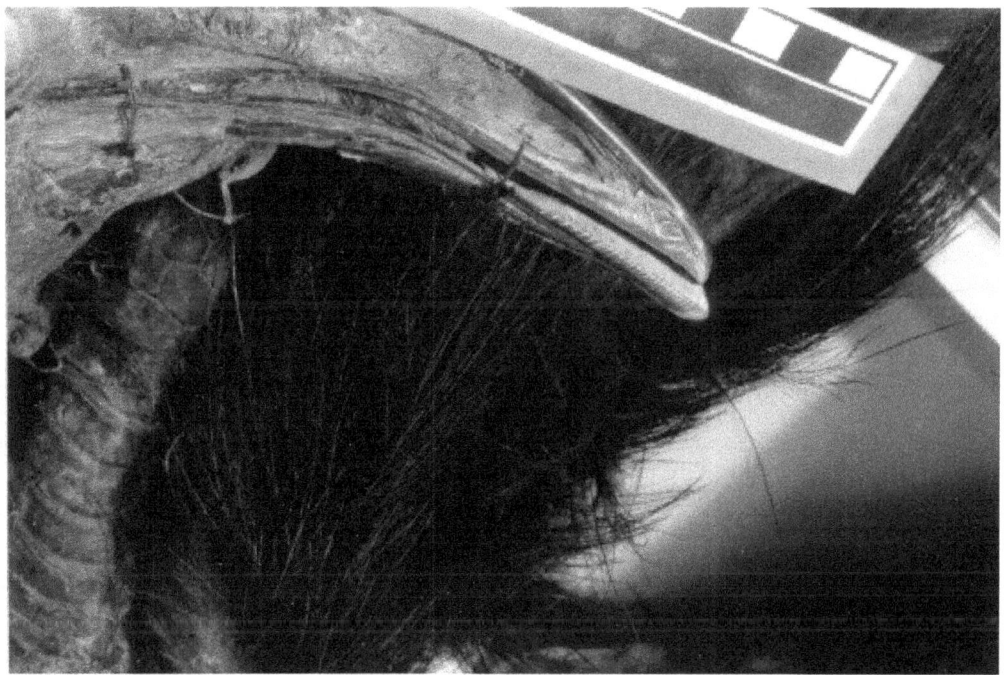

Fig. 6. Detail of *C. bennetti* jaw tips. Note the serrated border to the lower mandibular tomium. This specimen, a juvenile female, was collected in 1912 from New Guinea and was initially referred to the painted-necked species *C. picticollis*.

4.1. Wings, legs and feet

While cassowary wings are much reduced and largely hidden from view beneath their shaggy plumage, they still possess between three and five remiges (feathers growing from the hand and arm bones). These have become reduced to sharp-tipped, stiff quills (which may be up to 38 cm long) and are normally held streamlined into the rest of the plumage. The carpometacarpus is small and consists of a single fused mass with a tiny distal phalanx. Cassowaries have no rectrices. There are three toes (there is no hallux) and, remarkably, the claw on digit II is elongated into a sharp, long and rather straight weapon that, in *C. casuarius*, can be 12 cm long and 3 cm wide at the base (Fig. 7). It is not curved or otherwise suited for gripping or slicing, nor is it suited for climbing. Seriemas, which have an enlarged curved claw on pedal digit II, employ these claws when climbing. Cassowaries do not climb and their primary use of this enlarged claw is as an offensive weapon. It is well known that they will kick or leap and kick at an enemy, and they are even able to leap and throw both feet together in unison. They are reputed to be able to leap as high as 1.5 m off the ground (Hanzák and Formánek 1977).

4.2. Vertebral column, ribs and sternum

In vertebral count, cassowaries are like emus and ostriches but unlike rheas; moreover the three cassowary species differ from one another in their exact vertebral counts, usually by a difference of only one vertebra per section of the series. There are 14-16 neck vertebrae; 19-22 dorsals; 12-14 sacrals; and 8-9 caudals. A complete description of cassowary vertebrae, and those of all other living ratites, is provided by Mivart (1877). In *Casuarius* there are five sternal ribs, the first of which joins either the fourth or fifth thoracic rib. The fifth rib is about twice the length of the first and may not reach the sternum (it does not in the specimens I have examined). The cassowary sternum is of a characteristic boat-like shape that has a convex ventral surface and deeply concave medial surface. It is more elongate that the sterna of other ratites and has no prominence on the ventral surface (unlike rheas). Another unusual cassowary feature is the deeply marked pit on the cranial median process located between the coracoidal grooves: in emus this pit is also evident, but is only a tiny, poorly developed notch. Uncinate processes on the ribs are variably present in cassowaries - there may be as many as three of them on each side, or they may be entirely absent (Fig. 8).

5. Classifying cassowaries

Cassowaries have traditionally been regarded as distinct enough from all other ratites to deserve their own family, the Casuariidae Kaup 1847. Together with emus (the family Dromaiidae Gray 1870), they have conventionally been grouped in the Order Casuariiformes Sclater 1880. Sibley and Ahlquist (1990), the two geneticists who have reclassified all the living birds of the world based on DNA-DNA hybridization, have suggested that cassowaries and emus are more closely related than traditionally thought, and that both should be treated as subfamilies of a new, expanded family Casuariidae. Because certain fossil species appear intermediate between emus and cassowaries, most palaeornithologists also regard emus and cassowaries as representing closely related subfamilies.

Which ratites the casuariiforms are most closely related to proves to be a rather more thorny issue, and in the numerous classificatory schemes that have been published, virtually every conceivable relationship has been proposed. Based on his analysis of the ratite vertebral column, St. George Mivart (1877) was among the first to propose that ratites consist of a rhea-ostrich group and a casuariiform, kiwi and moa group. Garrod (1874) had proposed a similar arrangement based on thigh muscles and other features. However, in his classic and influential

Fig. 7. Plantar surface of a *C. bennetti* right foot. This specimen (which died in August 1909) was kept in captivity for some time. This explains the excessive overgrowth and wear of its claws. Nevertheless it serves to illustrate the often-significant difference in length between the claw of digit II and those of digits III and IV.

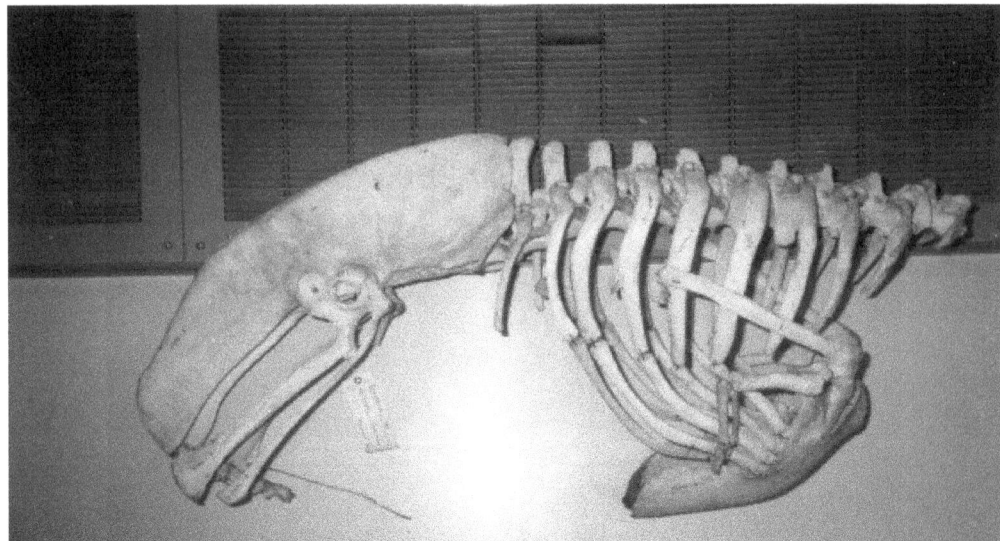

Fig. 8. At top, the thoracic, pectoral and pelvic skeleton of a cassowary (probably *C. casuarius*). Note the elongate, dorsoventrally shallow ilium, the small uncinate processes on the ribs and the deep sternum. Below, palmar view of a cassowary wing skeleton with feathers still attached. Note the proportionally tiny carpometacarpus and the elongate quills.

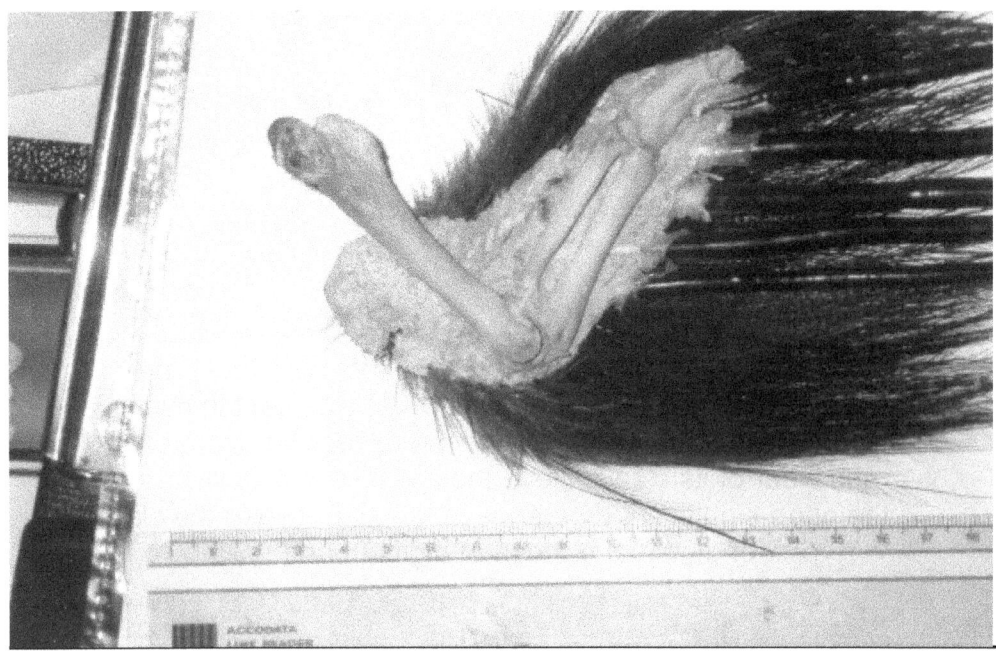

study of ratite evolution, Joel Cracraft (1974) concluded that casuariiforms were part of an assemblage, called the Struthiones, which also included elephant birds (Aepyornithidae), ostriches (Struthionidae) and rheas (Rheidae). Cracraft regarded casuariiforms as the sister-group to the ostrich-rhea clade (the Struthionoidea).

Studying the ratite tongue apparatus, Walter Bock and Paul Bühler (1988) argued for the opposite of Cracraft's phylogeny: that casuariiforms were part of an assemblage (the Tinami) that included tinamous, kiwis and rheas but excluded ostriches and elephant birds. Sibley and Ahlquist (1990) grouped the cassowary-emu group together with kiwis (Apterygidae) in a ratite group named Casuarii. The same grouping was supported by Anthony Bledsoe (1988) in his phylogenetic analysis of ratites based on features of the postcranial skeleton. In an analysis of both morphological and molecular characters, Lee et al. (1997) found that the morphological data supported Cracraft's (1974) tree, while the molecular data supported Sibley and Ahlquist's (1990) tree. They concluded that the morphological results were better supported. Most recently, in a study of ratite mitochondrial DNA, Marcel van Tuinen and colleagues (1998) found casuariiforms to be closest to rheas! In short, there are a plethora of opinions on ratite interrelations - the closest relatives of casuariiforms have been variously regarded as kiwis, ostriches or rheas (in other words, representing all the possible evolutionary combinations). New fossil discoveries and further investigations of ratite morphology and genetics will no doubt fuel further debate in the future.

6. Fossil cassowaries

DNA and fossil evidence indicate that cassowaries share an ancestor with emus that lived about 25 million years ago during the late Oligocene. Fossils regarded as emu have been reported from the Miocene of South Australia, and if cassowaries and emus are sister-groups, cassowaries must also have appeared by this time. There is therefore no reason to regard cassowaries as a particularly ancient group and correspondingly they do not have a very extensive fossil record. On the basis of their morphology and distribution, it seems that cassowaries are more primitive than emus and, in terms of proportions at least, it can be inferred that primitive emus looked more like cassowaries, rather than vice versa.

The most primitive known casuariid may be neither an emu nor a cassowary. Named *Dromaius gidju* by Chris Patterson and Patricia Vickers-Rich in 1987, this is an Oligocene-Miocene form that was originally described as a primitive emu. Discovered in the Wipajiri Formation's Kutjamurpu Local Fauna of South Australia, *D. gidju* was a small, gracile form that would have been about a metre tall. Patterson and Rich (1987) noted that *D. gidju* was less cursorial than other emus and suggested that future discoveries might demonstrate its distinction from true emus. This proved rather prophetic, as new material from Riversleigh in Queensland later allowed a reanalysis of the species. Published in 1992 by Walter Boles, this new material showed that *D. gidju* was something like an intermediate between cassowaries and emus: indeed, Boles has vernacularly referred to *D. gidju* as an 'cassomu' or an 'emuwary' (Archer et al. 1996). It was therefore regarded by Boles as worthy of a new genus and, in recognition of its combination of *Dromaius* and *Casuarius* feature, was named *Emuar*

Fig. 9. Double-wattled cassowary (*C. casuarius*).

ius (Boles 1992). While the feet and lower legs of *Emuarius* resemblehose of emus, its thigh and skull bones are like those of cassowaries.

These features indicate that *Emuarius* was less cursorial than modern emus but more cursorial than modern cassowaries.

Among true cassowary fossils are three pedal phalanges reported by Michael Plane (1967) from the Awe Fauna of southeast Papua New Guinea. This is of Pliocene age and is among the oldest known material identified as casuariine, and while the phalanges are more like those of *C. bennetti* than of other cassowary species, they are different enough to suggest that they do not belong to this species. Some toe bones from the Pliocene of New Guinea have also been referred to *Casuarius* by Vickers-Rich and van Tets (1982). The pedal phalanges of cassowaries are less dorsoventrally compressed than those of emus.

The best known fossil cassowary is *Casuarius lydekkeri*, a Pleistocene species based on a single partial right tibiotarsus first reported by Richard Lydekker (1891) and later named by Walter Rothschild (1911). As discussed at length by Vickers-Rich and colleagues in 1988, though this specimen has conventionally been regarded as having been discovered in the Wellington Caves of New South Wales (Miller 1962), its style of preservation does not match other fossils from this locality and Vickers-Rich and colleagues speculated that it may have been collected in New Guinea. What might be a second representative of this species was described from late Pleistocene deposits of Pureni, Papua New Guinea by Vickers-Rich et al. (1988). The tibiotarsus of the Pureni cassowary is a close match for that of *C. lydekkeri*, and Vickers-Rich et al. therefore chose not to name the Pureni specimen as a new species but regard it as another *C. lydekkeri* specimen. The Pureni cassowary, while fully adult, was very small – smaller than *C. bennetti* and perhaps less than a metre tall when alive. The Pureni pygmy differs notably from other cassowaries in having a narrower, shallower pelvis than any of the others and in having slimmer femora. It exhibits similarities with both *C. bennetti* and *C. casuarius* and thus how it is related to the other cassowary species remains uncertain.

7. **Cassowaries and people**

Cassowaries have proven to be important, both in the spiritual sense and the material sense, to the people who inhabit the same areas and, not only are the birds a valuable commodity to these people, they are also integral and powerful components of their belief systems.

As noted earlier, cassowaries have 'reversed' gender roles in that females are larger and more brightly coloured than males and not responsible for the care of juveniles. Female cassowaries are also apparently male 'mimics' in possessing a penis-like structure. These characteristics were certainly not missed by the Sambia and Mianmin people: regarding cassowaries as 'masculinised females', they believe that all cassowaries are biologically female yet equipped with male traits (similar views about emus are held by some Australian aboriginal tribes). In the ceremonies of the Umeda people, male dancers imitating cassowaries combine obvious masculine imagery (e.g., large penis gourds) with etiquette otherwise restricted to female dancers (e.g., holding hands). To the Bimin-Kuskusmin people who inhabit the central highlands of New Guinea, cassowaries represent the central character, named Afek, in a religious hierarchy where all the creatures are androgynous. Afek's brother/son/consort is either a fruit bat or an echidna named Yomnok, while both descend from an androgynous goanna. Cassowaries are also regarded as capable of reproducing without mating according to the legends of some New Guinea peoples and Shuker (1996) made the speculative but intriguing suggestion that this might hint at the possibility of parthenogenesis in these birds (reported for turkeys, so theoretically not impossible among other birds).

Recognizing that these peculiar ratites differ from 'normal' birds, aboriginal peoples also have some interesting taxonomical beliefs about the place of cassowaries in the natural world. Some tribes regard cassowaries as part of the same group as humans: after all, both are large, erect bipeds. Other tribes classify cassowaries with ordinary mammals on the basis of their hairy pelage, hard, bony skulls and terrestriality. The Waris and Arapesh people even regard cassowaries as capable of lactating - the young suckling from the pendant wattles or from the wing quills.

Nearly all of the information on aboriginal beliefs given here is based on information in Bagemihl (1999) - numerous supplementary references on aboriginal beliefs and legends concerning cassowaries are cited in that volume, see also Shuker (1996).

Despite their aggressive nature, cassowaries of all three species have long been kept in captivity by the people of New Guinea and Davies (1987) writes that they have been items of trade for at least 500 years. Their feathers have been used as decorations for head dresses and their stiff quills as nose ornaments. Cassowaries are also eaten and are raised to maturity before being slaughtered. Cassowary eggs are sometimes taken from clutches and the hatchling is then raised by a person whom it imprints on and follows around. Such birds reportedly become quite affectionate (Berridge 1934).

7.1. The scientific discovery of cassowaries

It may be surprising to find that European scientists knew about cassowaries long before they knew of emus. What might be regarded as 'the original cassowary' was a specimen initially collected on Banda Island (in the Moluccas) in or prior to 1595. This bird was bought, alive, to Amsterdam from Java in 1597 by a merchant company sent out to bring spices and merchandise from the Indonesian region. The Banda bird was then exhibited to the Dutch public and later changed hands several times, eventually becoming the property of Emperor Rudolph the Second. It was much figured and discussed in the literature of the 1600s and 1700s - Linneaus cited much of this literature in his work of 1758 and created the name *Struthio casuarius* for this species, citing its home as 'Asia, Sumatra, Molucca, Banda'. In 1760, Brisson recognised that the cassowary deserved its own genus and erected *Casuarius* based on the Papuan name. Following the publicity of the Banda specimen, cassowaries became so well known that emus were actually called 'New Holland cassowaries' until late in the 19th century.

It is also surprising that cassowaries from Banda Island remained the only known representatives of their kind until as late as 1854. In that year the naturalist Thomas Wall shot a cassowary near Cape York, Queensland, while on an expedition commanded by a Mr Kennedy. As described by Wall's brother (William Sheridan Wall) in an 1854 article in the *Illustrated Sydney Herald*, the flesh of this bird was eaten and its skin kept in a canvas bag. While this was unfortunately lost at Weymouth Bay (Thomas Wall himself later died on the same expedition), it seems that Wall's specimen was the first representative of *C. casuarius johnsoni*, a name created by Sclater in 1867. William Wall had actually proposed the name *Casuarius australis* for the Cape York specimen in the 1854 article but Sclater's name has become favoured.

Other kinds of cassowaries were named soon after the publication on the Cape York bird. Cassowaries from the Aru Islands, probably members of *C. casuarius*, were reported by the famous biogeographer Alfred Russel Wallace in 1857. This was also the year in which *C. bennetti* was first named by John Gould, this time for a specimen from New Britain. As with so many other scientifically important cassowary specimens, this one became well travelled and had been taken to Sydney by Captain Deolin in the cutter 'Oberon' (Rothschild 1900). Here it was examined by George Bennett who passed on his description of the bird to Gould. Bennett later sent the bird to London where, in 1863, it produced a clutch of eggs (though none of the young survived). The third of the presently recognized cassowary species, *C. unappendiculatus*, was named by Edward Blyth in 1860 for a juvenile specimen kept in the collections of the Bábu Rajendra Mullick of Calcutta. A second specimen of this species was present in the Zoological Garden at Amsterdam by 1871 and was apparently collected from the island of Salawati near New Guinea. Trade and the maintenance of these superb birds in captivity have therefore both been important factors in their scientific discovery and it is remarkable that, in the 1500s or earlier, live cassowaries were being shipped as far afield as Java and Europe. Anecdotal evidence of even earlier trade comes from a bas-relief depiction of a probable cassowary seen in a Javan Hindu temple (von Koenigswald and Steinbacher 1986, Shuker 1996).

Cassowaries have a small part to play in the 19th century reconstruction of the dinosaur *Iguanodon*. Confronted in 1882 with complete *Iguanodon* skeletons discovered in the Saint Barbe coal mine near Bernissart, Belgium, and kept at the Musée Royal d'Histoire Naturelle de Belgique, Louis Antoine Marie Joseph Dollo (1857-1931) used cassowary and wallaby skeletons as guides. Consequently he reconstructed *Iguanodon* as bipedal and erect bodied (a pose which more recent discoveries have shown to be incorrect). The cassowary skeleton seen in some of Dollo's photographs has often been incorrectly referred to as an emu.

7.2. The conservation of cassowaries

As denizens of a rainforest environment that faces destruction via logging practices, cassowaries are under threat and depend on forests with a rich diversity of fruiting species. As large and territorial animals that need healthy primary forest, their survival relies on the conservation of rainforest habitat. In Australia, the native subspecies *C. casuarius johnsonii* is listed as endangered and is regarded as a priority species for captive collections, meaning that its management and breeding in zoos and conservation parks are an integral part of its maintenance in captivity. A studbook of captive cassowaries is kept in Australia and there are plans for a release program in the future.

7.3. When cassowaries attack

It is well known that cassowaries can be dangerous, and indeed together with ostriches they are the only birds known to have definitely killed humans (see Naish 1999). On mainland Australia, the most recent recorded fatality occurred in April 1926 when Phillip McClean received an injury to the throat after running from a cassowary and falling to the ground. Attacking cassowaries charge and kick, sometimes jumping on top of the victim. Unlike emus, which reputedly kick backwards, cassowaries can kick in a forward and downward direction. They may also peck, barge or head-butt. The commonest injuries they cause in humans are puncture wounds, lacerations and bone fractures. Serious injuries resulting from cassowary attacks are most likely to occur if the person is crouching or is lying or has fallen on the ground. When confronted with a charging cassowary it is clearly unwise to crouch or turn one's back on the bird. Claims that jogging incites cassowaries to attack (supposedly because the sound of running feet imitates a foot-stamping rival cassowary) do not withstand scrutiny as walking people have been attacked more frequently than joggers.

However, cassowaries do not attack indiscriminately and a recent study by Christopher Kofron (1999) of 221 recorded attacks by *C. casuarius johnsonii* showed that attacks are mostly due to association of humans with food. Several attacks (7) appeared to be a territorial reaction to the presence of humans in an area where the cassowary was feeding while some (32) were clearly defensive – the cassowary either protecting itself or its chicks or eggs. McClean's death in 1926 was not the result of an unprovoked attack: he had struck the bird with the intention of killing it and had then fled. By far the greatest number of attacks (109) involved soliciting of food by the cassowary. In areas where humans have taken to feeding cassowaries, some cassowaries act boldly and aggressively in expectation of being fed and will run up to or chase people, sometimes kicking if no food is offered. Kofron reports that such behaviour was not recorded in his study area prior to 1985. Human feeding would thus appear to have modified cassowary behaviour and in fact cassowaries are naturally wary and highly unlikely to attack without provocation.

Cassowaries will also kick or peck at doors at windows, sometime breaking panes of glass or screen panels. In these cases they are presumably attacking a reflection, which they perceive as another cassowary. They will also kick or chase cars, again because they appear to associate the human occupants with food. Cassowaries dislike dogs and will attack them without provocation, presumably because feral dogs and dingos often prey on cassowaries. Between June 1996 and February 1997, six cassowaries were killed by dogs in the Cairns area and, of 35 cassowary attacks recorded by Kofron on dogs, 29 were in self-defence. Cassowaries also dislike cats. Attacks on horses and cows have also been recorded and *C. casuarius* is anecdotally credited with having killed small horses (C. Walker pers. comm.). These attacks were presumably territorial in motive.

7.4. Walter Rothschild and the Tring cassowaries

Lord Walter Rothschild (1868-1937) was quite probably the most important and prolific collector of zoological specimens during the late 19th and early 20th centuries. Working from his museum and home at Tring, Hertfordshire (still the site of both his museum and the BMNH bird collections), he amassed an unparalleled collection of literally thousands of insects, bird eggs, bird skins, mammal skins and other specimens. Rothschild appears to have been specially, if not almost fanatically, interested in a number of very specific groups of animals, among which were birds of paradise (see Fuller 1995), fleas, and cassowaries.

Virtually all of the information on Rothschild discussed here is taken from Miriam Rothschild's biography *Dear Lord Rothschild* (1983, ISI Press, Philadelphia).

Amassing one of the biggest single collections of cassowaries in the world, Rothschild's collection included no fewer than 62 **mounted** cassowaries. These specimens evidently prove rather problematic for the curators at Tring and Whitehead and Keates (1981) write '...for some reason Lord Rothschild decided to have no less than 65 [sic] of these large cassowaries mounted as if for future exhibition, and as such they make a unique collection and something of a headache for the curator'. Don't forget that this collection is augmented by many skins and skeletal specimens! Rothschild demanded that special attention be given to the mounting of cassowaries, and he only regarded one taxidermist - a man named Doggett - as able to complete the task with satisfactory results. Doggett was paid 30 pounds per cassowary mount by Rothschild; a sum regarded as extraordinarily high by Rothschild's curators and relatives and consequently curtailed in 1908 by Charles Rothschild, Walter's brother. The more than 60 mounted specimens eventually cost over 2000 pounds.
Studying these specimens with the aim of producing a monograph describing the different forms, Rothschild regarded it as essential that his descriptions were based on *live* specimens, not just on skins, so he collected all the specimens he could to keep and observe. Little has been published on how Rothschild and his staff maintained the birds, but it is known that they were not kept in a tropical house, nor heated at all. He once wrote, "My laying female has lived through 6 English winters without heat" (Rothschild 1983, p. 103). Given that cassowaries are famous for being pugnacious, one wonders if the cassowaries were ever the cause of any trouble. Indeed, cassowaries were partially responsible for the harsh attitude Rothschild's father (Nathaniel Rothschild) had of his son's collection for, in 1888, one of the cassowaries which roamed free in Tring Park attacked Nathaniel's horse. However, Rothschild did complete his work and, in 1900, published his definitive monograph on the cassowaries (Rothschild 1900). It is a lavishly illustrated work in which several new forms are named based on the colour of their necks or configuration of their wattles - features now regarded as too variable for much basis in taxonomy.

Though it might seem that Rothschild's work on cassowaries had now reached fruition, he continued to collect the birds and started to amass a secret collection of live specimens. While his father was prohibiting the further purchase of specimens, Rothschild wrote to his collectors to continue shipping live cassowaries, but to keep them at a safe location, rather than send them straight to Tring.

Rothschild had a dark secret: he was being blackmailed by a wealthy aristocratic former mistress. Aided by her husband, this woman eventually forced Rothschild, in 1931, to sell the better part of his ornithological collection to the American Museum of Natural History for $225,000 - about a dollar a specimen.

The blackmailer remains anonymous but Miriam Rothschild (1983) states that she is aware of her true identity. Despite this tremendous and devastating loss (it is known that Rothschild was, understandably, very upset by these proceedings), Rothschild could not part with his cassowaries and all of the specimens - the mounts, skins and skeletons - were retained at Tring. Today they represent an invaluable collection with a fascinating history.

7.5. Cassowaries in captivity

Cassowaries have not proved easy to breed in captivity. While chicks were hatched as early as 1862 and 1863 (both at London Zoo), these did not survive and the first successful hatching and rearing of a chick appears to be from 1957 when a single *C. casuarius* chick was raised at San Diego Zoo. The male parent of this chick was a long-time resident of the zoo and had been there for 31 years. In recent decades several collections around the world have had success in hatching and raising cassowaries. Among the more notable have been Edinburgh Zoo (Scotland), Airlie Beach Wildlife Park (Australia) and Denver Zoo (USA). Airlie Beach has proved Australia's most prolific cassowary breeding collection, having produced 22 birds to date, while Denver has the world record – between 1977 and 1992 98 chicks were bred.

It is not advised that captive cassowaries be kept with other species as they may attack and kill them. At Currumbin Sanctuary, Queensland, a cassowary has killed an ibis and attacked some eastern wallaroos that entered its enclosure. However, cassowaries have also been successfully housed with other animals without incident.

References

- Archer, M. Hand, S. J. and Godthelp, H. 1996. *Riversleigh*. Reed Books (Victoria).
- Bagemihl, B. 1999. *Biological Exuberance: Animal Homosexuality and Natural Diversity*. Profile Books (London).
- Berridge, W. S. 1934. *All About Birds*. Harrap (London).
- Bledsoe, A. H. 1988. A phylogenetic analysis of postcranial skeletal characters of the ratite birds. *Annals of Carnegie Museum* **57**: 73-90.
- Bock, W. J. and Buhler, P. 1988. The evolution and biogeographical history of the paleognathous birds. *Current Topics in Avian Biology*. Proceedings of the One-hundredth Meeting, Deutsche Ornithologen-Gesellshaft, Bonn.

- Boles, W. E. 1992. Revision of *Dromaius gidju* Patterson and Rich 1987 from Riversleigh, northwestern Queensland, Australia, with a reassessment of its generic position. *Natural History Museum of Los Angeles County, Science Series* **36**: 195-208.
- Burton, M. (ed) 1984. *Encyclopaedia of Animals*. Cathay (London).
- Cracraft, J. 1974. Phylogeny and evolution of ratite birds. *Ibis* **115**: 494-521.
- Cree, A. & Daugherty, C. 1990. Tuatara sheds its fossil image. *New Scientist* **128** (1739): 30-34.
- Crome, F. H. J. 1975. Some observations on the biology of the cassowary in northern Queensland. *Emu* **76**: 8-14.
- Davies, S. J. J. F. 1987. Cassowaries. IN Perrins, C. M. and Middleton, A. L. A. (eds) *Animals of the World: Water Birds and Flightless Birds*. Leisure Circle (Wembley, UK), pp. 21-23.
- Davies, S. J. J. F. 1991. Ratites and tinamous. IN Forshaw, J. (ed) *Encyclopedia of Animals: Birds*. Merehurst (London), pp. 46-49.
- Durrell, G. 1977. *Golden bats and pink pigeons: a journey to the flora and fauna of a unique island*. Simon and Schuster (New York).
- Fuller, E. 1995. *The Lost Birds of Paradise*. Swan Hill Press (Shrewsbury, UK).
- Garrod, A. H. 1874. On certain muscles of the thigh of birds and on their value in classification. Part II. *Proceedings of the Zoological Society of London* **1874**: 111-123.
- Hanzák, J. and Formánek, J. 1977. *The Illustrated Encyclopedia of Birds*. Octopus (London).
- Koenigswald, G. H. R. von and Steinbacher, J. 1986. Fremde vögel an fernem ort. *Natur und Museum* **116**: 97-103.
- Kofron, C. P. 1999. Attacks to humans and domestic animals by the southern cassowary (*Casuarius casuarius johnsonii*) in Queensland, Australia. *Journal of Zoology* **249**: 375-381.
- Lee, K., Feinstein, J. and Cracraft, J. 1997. The phylogeny of ratite birds: resolving conflicts between molecular and morphological data sets. IN Mindell, D. P. (ed) *Avian Molecular Evolution and Systematics*. Academic Press (London), pp. 173-211.
- Lydekker, R. 1891. *Catalogue of the Fossil Birds in the British Museum (Natural History)*. British Museum (London).
- Miller, A. H. 1962. The history and significance of the fossil *Casuarius lydekkeri*. *Records of the Australian Museum* **25**: 235-238.
- Mivart, St. G. 1877. On the axial skeleton of the Struthionidae. *Transactions of the Zoological Society, London* **10**: 1-52.
- Naish D. 1999. Big bad killer eagles. *Fortean Times* **122**: 48.
- Owadally, A. W. 1979. The dodo and the tambalacoque tree. *Science* **203**: 1363-1364.
- Patterson, C. and Vickers-Rich, P. 1987. The fossil history of the emus, *Dromaius* (Aves: Dromaiinae). *Records of the South Australian Museum* **36**: 63-126.
- Plane, M. D. 1967. Stratigraphy and vertebrate fauna of the Otibanda Formation, New Guinea. *Bureau of Mineral Resources, Geology and Geophysics (Australia) Bulletin* **184**: 1-64.
- Rothschild, W. 1911. On the former and present distribution of the so-called Ratitae or ostrich-like birds. IN *Verh. V International Ornithological Congress, Berlin*, pp. 144-169.

- Rothschild, M. 1983. *Dear Lord Rothschild: Birds, butterflies and history*. Balaban Publishers (Glenside, Pennsylvania).
- Rothschild, W. 1900. A monograph of the genus *Casuarius*. *Transactions of the Zoological Society, London* **15**: 109-148.
- Sibley, C. G. and Ahlquist, J. E. 1990. *Phylogeny and classification of the birds: a study in molecular evolution*. Yale Univ. Press (New Haven and London).
- Shuker, K. P. N. 1996. Super-emus and spinifex men. IN Moore, S. (ed) *Fortean Studies Volume 3*. John Brown Publishing (London), pp. 189-210.
- Shuker, K. P. N. 1997. Aardvark anomalies. IN Downes, S. (ed) *The CFZ Yearbook 1997*. CFZ Publications (Exeter), pp. 116-134.
- Stocker G and Irvine A. 1983. Seed dispersal by cassowaries (*Casuarius casuarius*) in north Queensland's rainforests. *Biotropica* **15**: 170-176.
- Temple, S. A. 1977. Plant-animal mutualism: coevolution with dodo leads to near extinction of plant. *Science* **197**: 885-886.
- Van Tuinen, M., Sibley, C. G. and Hedges, S. B. 1998. Phylogeny and biogeography of ratite birds inferred from DNA sequences of the mitochondrial ribosomal genes. *Molecular Biological Evolution* **15**: 370-376.
- Vickers-Rich, P. 1991. The Mesozoic and Tertiary history of birds on the Australian plate. IN Vickers-Rich, P., Monaghan, T. M., Baird, R. R. and Rich, T. H. *Vertebrate Palaeontology of Australasia*. Pioneer Design Studio & Monash Univ. (Melbourne, Aus.), pp. 721-808.
- Vickers-Rich, P., Plane, M. D. and Schroeder, N. 1988. A pygmy cassowary (*Casuarius lydekkeri*) from late Pleistocene bog deposits at Pureni, Papua New Guina. *Bureau of Mineral Resources Journal of Australian Geology and Geophysics* **10**: 377-389.
- Vickers-Rich, P. and Van Tets, G. F. 1982. Fossil birds of Australia and New Guinea: their biogeographic, phylogenetic and biostratigraphic input. IN Vickers-Rich, P. and Thompson, E. M. (eds) *The Fossil Vertebrate Record of Australia*. Monash University (Clayton, Aus.), pp. 235-384.
- White, C. M. N. 1975. The problem of the cassowary in Ceram. *Bulletin of the British Ornithologists' Club* **95**: 165-170.
- White, C. M. N. 1976. The problem of the cassowary in New Britain. *Bulletin of the British Ornithologists' Club* **96**: 66-68.
- Whitehead, P. J. P. and Keates, C. 1981. *The British Museum (Natural History)*. P. Wilson (London).
- Witmer, M. C. and Cheke, A. S. 1991. The dodo and the tambalocoque tree: an obligate mutualism reconsidered. *Oikis* **61**: 133-137.

THE CENTRE FOR FORTEAN ZOOLOGY

So, what is the Centre for Fortean Zoology?

We are a non profit-making organisation founded in 1992 with the aim of being a clearing house for information, and coordinating research into mystery animals around the world. We also study out of place animals, rare and aberrant animal behaviour, and Zooform Phenomena; little-understood "things" that appear to be animals, but which are in fact nothing of the sort, and not even alive (at least in the way we understand the term).

Why should I join the Centre for Fortean Zoology?

Not only are we the biggest organisation of our type in the world, but - or so we like to think - we are the best. We are certainly the only truly global Cryptozoological research organisation, and we carry out our investigations using a strictly scientific set of guidelines. We are expanding all the time and looking to recruit new members to help us in our research into mysterious animals and strange creatures across the globe. Why should you join us? Because, if you are genuinely interested in trying to solve the last great mysteries of Mother Nature, there is nobody better than us with whom to do it.

What do I get if I join the Centre for Fortean Zoology?

For £12 a year, you get a four issue subscription to our journal *Animals & Men*. Each issue contains 60 pages packed with news, articles, letters, research papers, field reports, and even a gossip column! The magazine is A5 in format with a full colour cover. You also have access to one of the world's largest collections of resource material dealing with cryptozoology and allied disciplines, and people from the CFZ membership regularly take part in fieldwork and expeditions around the world.

How is the Centre for Fortean Zoology organized?

The CFZ is managed by a three-man board of trustees, with a non-profit making trust registered with HM Government Stamp Office. The board of trustees is supported by a Permanent Directorate of full and part-time staff, and advised by a Consultancy Board of specialists - many of whom who are world-renowned experts in their particular field. We have regional representatives across the UK, the USA, and many other parts of the world, and are affiliated with other organisations whose aims and protocols mirror our own.

I am new to the subject, and although I am interested I have little practical knowledge. I don't want to feel out of my depth. What should I do?

Don't worry. We were *all* beginners once. You'll find that the people at the CFZ are friendly and approachable. We have a thriving forum on the website which is the hub of an ever-growing electronic community. You will soon find your feet. Many members of the CFZ Permanent Directorate started off as ordinary members, and now work full-time chasing monsters around the world.

I have an idea for a project which isn't on your website. What do I do?

Write to us, e-mail us, or telephone us. The list of future projects on the website is not exhaustive. If you have a good idea for an investigation, please tell us. We may well be able to help.

How do I go on an expedition?

We are always looking for volunteers to join us. If you see a project that interests you, do not hesitate to get in touch with us. Under certain circumstances we can help provide funding for your trip. If you look on the future projects section of the website, you can see some of the projects that we have pencilled in for the next few years.

In 2003 and 2004 we sent three-man expeditions to Sumatra looking for Orang-Pendek - a semi-legendary bipedal ape. The same three went to Mongolia in 2005. All three members started off merely subscribers to the CFZ magazine.

Next time it could be you!

Project Kerinci, Sumatra - 2003
In search of the bipedal ape Orang Pendek

How is the Centre for Fortean Zoology funded?

We have no magic sources of income. All our funds come from donations, membership fees, works that we do for TV, radio or magazines, and sales of our publications and merchandise. We are always looking for corporate sponsorship, and other sources of revenue. If you have any ideas for fund-raising please let us know. However, unlike other cryptozoological organisations in the past, we do not live in an intellectual ivory tower. We are not afraid to get our hands dirty, and furthermore we are not one of those organisations where the membership have to raise money so that a privileged few can go on expensive foreign trips. Our research teams both in the UK and abroad, consist of a mixture of experienced and inexperienced personnel. We are truly a community, and work on the premise that the benefits of CFZ membership are open to all.

What do you do with the data you gather from your investigations and expeditions?

Reports of our investigations are published on our website as soon as they are available. Preliminary reports are posted within days of the project finishing.

Each year we publish a 200 page yearbook containing research papers and expedition reports too long to be printed in the journal. We freely circulate our information to anybody who asks for it.

Is the CFZ community purely an electronic one?

No. Each year since 2000 we have held our annual convention - the *Weird Weekend* - in Exeter. It is three days of lectures, workshops, and excursions. But most importantly it is a chance for members of the CFZ to meet each other, and to talk with the members of the permanent directorate in a relaxed and informal setting and preferably with a pint of beer in one hand. Since 2006 - the *Weird Weekend* has been bigger and better and held in the idyllic rural location of Woolsery in North Devon. The 2008 event will be held over the weekend 15-17 August.

Since relocating to North Devon in 2005 we have become ever more closely involved with other community organisations, and we hope that this trend will continue. We also work closely with Police Forces across the UK as consultants for animal mutilation cases, and we intend to forge closer links with the coastguard and other community services. We want to work closely with those who regularly travel into the Bristol Channel, so that if the recent trend of exotic animal visitors to our coastal waters continues, we can be out there as soon as possible.

We are building a Visitor's Centre in rural North Devon. This will not be open to the general public, but will provide a museum, a library and an educational resource for our members (currently over 400) across the globe. We are also planning a youth organisation which will involve children and young people in our activities. We work closely with *Tropiquaria* - a small zoo in north Somerset, and have several exciting conservation projects planned.

Apart from having been the only Fortean Zoological organisation in the world to have consistently published material on all aspects of the subject for over a decade, we have achieved the following concrete results:

- Disproved the myth relating to the headless so-called sea-serpent carcass of Durgan beach in Cornwall 1975
- Disproved the story of the 1988 puma skull of Lustleigh Cleave
- Carried out the only in-depth research ever into the mythos of the Cornish Owlman
- Made the first records of a tropical species of lamprey
- Made the first records of a luminous cave gnat larva in Thailand.
- Discovered a possible new species of British mammal - the beech marten.
- In 1994-6 carried out the first archival fortean zoological survey of Hong Kong.
- In the year 2000, CFZ theories where confirmed when an entirely new species of lizard was found resident in Britain.
- Identified the monster of Martin Mere in Lancashire as a giant wels catfish
- Expanded the known range of Armitage's skink in the Gambia by 80%
- Obtained photographic evidence of the remains of Europe's largest known pike
- Carried out the first ever in-depth study of the *ninki-nanka*
- Carried out the first attempt to breed Puerto Rican cave snails in captivity
- Were the first European explorers to visit the `lost valley` in Sumatra
- Published the first ever evidence for a new tribe of pygmies in Guyana
- Published the first evidence for a new species of caiman In Guyana

EXPEDITIONS & INVESTIGATIONS TO DATE INCLUDE:

- 1998 Puerto Rico, Florida, Mexico *(Chupacabras)*
- 1999 Nevada *(Bigfoot)*
- 2000 Thailand *(Giant snakes called nagas)*
- 2002 Martin Mere *(Giant catfish)*
- 2002 Cleveland *(Wallaby mutilation)*
- 2003 Bolam Lake *(BHM Reports)*
- 2003 Sumatra *(Orang Pendek)*
- 2003 Texas *(Bigfoot; giant snapping turtles)*
- 2004 Sumatra *(Orang Pendek; cigau, a sabre-toothed cat)*
- 2004 Illinois *(Black panthers; cicada swarm)*
- 2004 Texas *(Mystery blue dog)*
- 2004 Puerto Rico *(Chupacabras; carnivorous cave snails)*
- 2005 Belize *(Affiliate expedition for hairy dwarfs)*
- 2005 Mongolia *(Allghoi Khorkhoi aka Mongolian death worm)*
- 2006 Gambia *(Gambo - Gambian sea monster , Ninki Nanka and Armitage s skink*
- 2006 Llangorse Lake *(Giant pike, giant eels)*
- 2006 Windermere *(Giant eels)*
- 2007 Coniston Water *(Giant eels)*
- 2007 Guyana *(Giant anaconda, didi, water tiger)*

To apply for a <u>FREE</u> information pack about the organisation and details of how to join, plus information on current and future projects, expeditions and events.

Send a stamped and addressed envelope to:

**THE CENTRE FOR FORTEAN ZOOLOGY
MYRTLE COTTAGE, WOOLSERY,
BIDEFORD, NORTH DEVON
EX39 5QR.**

or alternatively visit our website at:
www.cfz.org.uk

Other books available from
CFZ PRESS

THE OWLMAN AND OTHERS - 30th Anniversary Edition
Jonathan Downes - ISBN 978-1-905723-02-7

£14.99

EASTER 1976 - Two young girls playing in the churchyard of Mawnan Old Church in southern Cornwall were frightened by what they described as a "nasty bird-man". A series of sightings that has continued to the present day. These grotesque and frightening episodes have fascinated researchers for three decades now, and one man has spent years collecting all the available evidence into a book. To mark the 30th anniversary of these sightings, Jonathan Downes has published a special edition of his book.

DRAGONS - More than a myth?
Richard Freeman - ISBN 0-9512872-9-X

£14.99

First scientific look at dragons since 1884. It looks at dragon legends worldwide, and examines modern sightings of dragon-like creatures, as well as some of the more esoteric theories surrounding dragonkind.

Dragons are discussed from a folkloric, historical and cryptozoological perspective, and Richard Freeman concludes that: "When your parents told you that dragons don't exist - they lied!"

MONSTER HUNTER
Jonathan Downes - ISBN 0-9512872-7-3

£14.99

Jonathan Downes' long awaited autobiography, *Monster Hunter*...

Written with refreshing candour, it is the extraordinary story of an extraordinary life, in which the author crosses paths with wizards, rock stars, terrorists, and a bewildering array of mythical and not so mythical monsters, and still just about manages to emerge with his sanity intact.......

MONSTER OF THE MERE
Jonathan Downes - ISBN 0-9512872-2-2

£12.50

It all starts on Valentine's Day 2002 when a Lancashire newspaper announces that "Something" has been attacking swans at a nature reserve in Lancashire. Eyewitnesses have reported that a giant unknown creature has been dragging fully grown swans beneath the water at Martin Mere. An intrepid team from the Exeter based Centre for Fortean Zoology, led by the author, make two trips – each of a week – to the lake and its surrounding marshlands. During their investigations they uncover a thrilling and complex web of historical fact and fancy, quasi Fortean occurrences, strange animals and even human sacrifice.

**CFZ PRESS, MYRTLE COTTAGE,
WOOLFARDISWORTHY BIDEFORD,
NORTH DEVON, EX39 5QR
www.cfz.org.uk**

Other books available from
CFZ PRESS

ONLY FOOLS AND GOATSUCKERS
Jonathan Downes - ISBN 0-9512872-3-0

£12.50

In January and February 1998 Jonathan Downes and Graham Inglis of the Centre for Fortean Zoology spent three and a half weeks in Puerto Rico, Mexico and Florida, accompanied by a film crew from UK Channel 4 TV. Their aim was to make a documentary about the terrifying chupacabra - a vampiric creature that exists somewhere in the grey area between folklore and reality. This remarkable book tells the gripping, sometimes scary, and often hilariously funny story of how the boys from the CFZ did their best to subvert the medium of contemporary TV documentary making and actually do their job.

WHILE THE CAT'S AWAY
Chris Moiser - ISBN: 0-9512872-1-4

£7.99

Over the past thirty years or so there have been numerous sightings of large exotic cats, including black leopards, pumas and lynx, in the South West of England. Former Rhodesian soldier Sam McCall moved to North Devon and became a farmer and pub owner when Rhodesia became Zimbabwe in 1980. Over the years despite many of his pub regulars having seen the "Beast of Exmoor" Sam wasn't at all sure that it existed. Then a series of happenings made him change his mind. Chris Moiser—a zoologist—is well known for his research into the mystery cats of the westcountry. This is his first novel.

CFZ EXPEDITION REPORT 2006 - GAMBIA
ISBN 1905723032

£12.50

In July 2006, The J.T.Downes memorial Gambia Expedition - a six-person team - Chris Moiser, Richard Freeman, Chris Clarke, Oll Lewis, Lisa Dowley and Suzi Marsh went to the Gambia, West Africa. They went in search of a dragon-like creature, known to the natives as `Ninki Nanka`, which has terrorized the tiny African state for generations, and has reportedly killed people as recently as the 1990s. They also went to dig up part of a beach where an amateur naturalist claims to have buried the carcass of a mysterious fifteen foot sea monster named 'Gambo', and they sought to find the Armitage's Skink (*Chalcides armitagei*) - a tiny lizard first described in 1922 and only rediscovered in 1989. Here, for the first time, is their story.... With an forward by Dr. Karl Shuker and introduction by Jonathan Downes.

BIG CATS IN BRITAIN YEARBOOK 2006
Edited by Mark Fraser - ISBN 978-1905723-01-0

£10.00

Big cats are said to roam the British Isles and Ireland even now as you are sitting and reading this. People from all walks of life encounter these mysterious felines on a daily basis in every nook and cranny of these two countries. Most are jet-black, some are white, some are brown, in fact big cats of every description and colour are seen by some unsuspecting person while on his or her daily business. 'Big Cats in Britain' are the largest and most active group in the British Isles and Ireland This is their first book. It contains a run-down of every known big cat sighting in the UK during 2005, together with essays by various luminaries of the British big cat research community which place the phenomenon into scientific, cultural, and historical perspective.

CFZ PRESS, MYRTLE COTTAGE,
WOOLSERY, BIDEFORD,
NORTH DEVON, EX39 5QR
w w w . c f z . o r g . u k

Other books available from
CFZ PRESS

THE SMALLER MYSTERY CARNIVORES OF THE WESTCOUNTRY
Jonathan Downes - ISBN 978-1-905723-05-8

£7.99

Although much has been written in recent years about the mystery big cats which have been reported stalking Westcountry moorlands, little has been written on the subject of the smaller British mystery carnivores. This unique book redresses the balance and examines the current status in the Westcountry of three species thought to be extinct: the Wildcat, the Pine Marten and the Polecat, finding that the truth is far more exciting than the currently held scientific dogma. This book also uncovers evidence suggesting that even more exotic species of small mammal may lurk hitherto unsuspected in the countryside of Devon, Cornwall, Somerset and Dorset.

THE BLACKDOWN MYSTERY
Jonathan Downes - ISBN 978-1-905723-00-3

£7.99

Intrepid members of the CFZ are up to the challenge, and manage to entangle themselves thoroughly in the bizarre trappings of this case. This is the soft underbelly of ufology, rife with unsavoury characters, plenty of drugs and booze." That sums it up quite well, we think. A new edition of the classic 1999 book by legendary fortean author Jonathan Downes. In this remarkable book, Jon weaves a complex tale of conspiracy, anti-conspiracy, quasi-conspiracy and downright lies surrounding an air-crash and alleged UFO incident in Somerset during 1996. However the story is much stranger than that. This excellent and amusing book lifts the lid off much of contemporary forteana and explains far more than it initially promises.

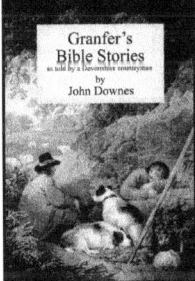

GRANFER'S BIBLE STORIES
John Downes - ISBN 0-9512872-8-1

£7.99

Bible stories in the Devonshire vernacular, each story being told by an old Devon Grandfather - 'Granfer'. These stories are now collected together in a remarkable book presenting selected parts of the Bible as one more-or-less continuous tale in short 'bite sized' stories intended for dipping into or even for bed-time reading. `Granfer` treats the biblical characters as if they were simple country folk living in the next village. Many of the stories are treated with a degree of bucolic humour and kindly irreverence, which not only gives the reader an opportunity to re-evaluate familiar tales in a new light, but do so in both an entertaining and a spiritually uplifting manner.

FRAGRANT HARBOURS DISTANT RIVERS
John Downes - ISBN 0-9512872-5-7

£12.50

Many excellent books have been written about Africa during the second half of the 19th Century, but this one is unique in that it presents the stories of a dozen different people, whose interlinked lives and achievements have as many nuances as any contemporary soap opera. It explains how the events in China and Hong Kong which surrounded the Opium Wars, intimately effected the events in Africa which take up the majority of this book. The author served in the Colonial Service in Nigeria and Hong Kong, during which he found himself following in the footsteps of one of the main characters in this book; Frederick Lugard – the architect of modern Nigeria.

**CFZ PRESS, MYRTLE COTTAGE,
WOOLFARDISWORTHY BIDEFORD,
NORTH DEVON, EX39 5QR
www.cfz.org.uk**

Other books available from
CFZ PRESS

ANIMALS & MEN - Issues 1 - 5 - In the Beginning
Edited by Jonathan Downes - ISBN 0-9512872-6-5

£12.50

At the beginning of the 21st Century monsters still roam the remote, and sometimes not so remote, corners of our planet. It is our job to search for them. The Centre for Fortean Zoology [CFZ] is the only professional, scientific and full-time organisation in the world dedicated to cryptozoology - the study of unknown animals. Since 1992 the CFZ has carried out an unparalleled programme of research and investigation all over the world. We have carried out expeditions to Sumatra (2003 and 2004), Mongolia (2005), Puerto Rico (1998 and 2004), Mexico (1998), Thailand (2000), Florida (1998), Nevada (1999 and 2003), Texas (2003 and 2004), and Illinois (2004). An introductory essay by Jonathan Downes, notes putting each issue into a historical perspective, and a history of the CFZ.

ANIMALS & MEN - Issues 6 - 10 - The Number of the Beast
Edited by Jonathan Downes - ISBN 978-1-905723-06-5

£12.50

At the beginning of the 21st Century monsters still roam the remote, and sometimes not so remote, corners of our planet. It is our job to search for them. The Centre for Fortean Zoology [CFZ] is the only professional, scientific and full-time organisation in the world dedicated to cryptozoology - the study of unknown animals. Since 1992 the CFZ has carried out an unparalleled programme of research and investigation all over the world. We have carried out expeditions to Sumatra (2003 and 2004), Mongolia (2005), Puerto Rico (1998 and 2004), Mexico (1998), Thailand (2000), Florida (1998), Nevada (1999 and 2003), Texas (2003 and 2004), and Illinois (2004). Preface by Mark North and an introductory essay by Jonathan Downes, notes putting each issue into a historical perspective, and a history of the CFZ.

BIG BIRD! Modern Sightings of Flying Monsters

£7.99

Ken Gerhard - ISBN 978-1-905723-08-9

From all over the dusty U.S./Mexican border come hair-raising stories of modern day encounters with winged monsters of immense size and terrifying appearance. Further field sightings of similar creatures are recorded from all around the globe. What lies behind these weird tales? Ken Gerhard is a native Texan, he lives in the homeland of the monster some call 'Big Bird'. Ken's scholarly work is the first of its kind. On the track of the monster, Ken uncovers cases of animal mutilations, attacks on humans and mounting evidence of a stunning zoological discovery ignored by mainstream science. Keep watching the skies!

STRENGTH THROUGH KOI
They saved Hitler's Koi and other stories

£7.99

Jonathan Downes - ISBN 978-1-905723-04-1

Strength through Koi is a book of short stories - some of them true, some of them less so - by noted cryptozoologist and raconteur Jonathan Downes. The stories are all about koi carp, and their interaction with bigfoot, UFOs, and Nazis. Even the late George Harrison makes an appearance. Very funny in parts, this book is highly recommended for anyone with even a passing interest in aquaculture, but should be taken definitely *cum grano salis*.

CFZ PRESS, MYRTLE COTTAGE, WOOLSERY, BIDEFORD, NORTH DEVON, EX39 5QR

Other books available from
CFZ PRESS

BIG CATS IN BRITAIN YEARBOOK 2007
Edited by Mark Fraser - ISBN 978-1-905723-09-6

£12.50

People from all walks of life encounter mysterious felids on a daily basis, in every nook and cranny of the UK. Most are jet-black, some are white, some are brown; big cats of every description and colour are seen by some unsuspecting person while on his or her daily business. 'Big Cats in Britain' are the largest and most active research group in the British Isles and Ireland. This book contains a run-down of every known big cat sighting in the UK during 2006, together with essays by various luminaries of the British big cat research community.

CAT FLAPS! Northern Mystery Cats
Andy Roberts - ISBN 978-1-905723-11-9

£6.99

Of all Britain's mystery beasts, the alien big cats are the most renowned. In recent years the notoriety of these uncatchable, out-of-place predators have eclipsed even the Loch Ness Monster. They slink from the shadows to terrorise a community, and then, as often as not, vanish like ghosts. But now film, photographs, livestock kills, and paw prints show that we can no longer deny the existence of these once-legendary beasts. Here then is a case-study, a true lost classic of Fortean research by one of the country's most respected researchers.

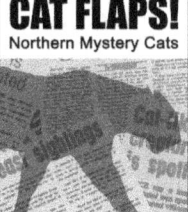

CENTRE FOR FORTEAN ZOOLOGY 2007 YEARBOOK
Edited by Jonathan Downes and Richard Freeman
ISBN 978-1-905723-14-0

£12.50

The Centre For Fortean Zoology Yearbook is a collection of papers and essays too long and detailed for publication in the CFZ Journal *Animals & Men*. With contributions from both well-known researchers, and relative newcomers to the field, the Yearbook provides a forum where new theories can be expounded, and work on little-known cryptids discussed.

MONSTER! THE A-Z OF ZOOFORM PHENOMENA
Neil Arnold - ISBN 978-1-905723-10-2

£14.99

Zooform Phenomena are the most elusive, and least understood, mystery `animals`. Indeed, they are not animals at all, and are not even animate in the accepted terms of the word. Author and researcher Neil Arnold is to be commended for a groundbreaking piece of work, and has provided the world's first alphabetical listing of zooforms from around the world.

**CFZ PRESS, MYRTLE COTTAGE,
WOOLFARDISWORTHY BIDEFORD,
NORTH DEVON, EX39 5QR
w w w . c f z . o r g . u k**

Other books available from
CFZ PRESS

BIG CATS LOOSE IN BRITAIN
Marcus Matthews - ISBN 978-1-905723-12-6

£14.99

Big Cats: Loose in Britain, looks at the body of anecdotal evidence for such creatures: sightings, livestock kills, paw-prints and photographs, and seeks to determine underlying commonalities and threads of evidence. These two strands are repeatedly woven together into a highly readable, yet scientifically compelling, overview of the big cat phenomenon in Britain.

DARK DORSET
TALES OF MYSTERY, WONDER AND TERROR
Robert. J. Newland and Mark. J. North
ISBN 978-1-905723-15-6

£12.50

This extensively illustrated compendium has over 400 tales and references, making this book by far one of the best in its field. Dark Dorset has been thoroughly researched, and includes many new entries and up to date information never before published. The title of the book speaks for itself, and is indeed not for the faint hearted or those easily shocked.

MAN-MONKEY - IN SEARCH OF THE BRITISH BIGFOOT
Nick Redfern - ISBN 978-1-905723-16-4

£9.99

In her 1883 book, *Shropshire Folklore*, Charlotte S. Burne wrote: *'Just before he reached the canal bridge, a strange black creature with great white eyes sprang out of the plantation by the roadside and alighted on his horse's back'*. The creature duly became known as the `Man-Monkey`.

Between 1986 and early 2001, Nick Redfern delved deeply into the mystery of the strange creature of that dark stretch of canal. Now, published for the very first time, are Nick's original interview notes, his files and discoveries; as well as his theories pertaining to what lies at the heart of this diabolical legend.

EXTRAORDINARY ANIMALS REVISITED
Dr Karl Shuker - ISBN 978-1905723171

£14.99

This delightful book is the long-awaited, greatly-expanded new edition of one of Dr Karl Shuker's much-loved early volumes, *Extraordinary Animals Worldwide*. It is a fascinating celebration of what used to be called romantic natural history, examining a dazzling diversity of animal anomalies, creatures of cryptozoology, and all manner of other thought-provoking zoological revelations and continuing controversies down through the ages of wildlife discovery.

**CFZ PRESS, MYRTLE COTTAGE,
WOOLFARDISWORTHY BIDEFORD,
NORTH DEVON, EX39 5QR
w w w . c f z . o r g . u k**

Other books available from
CFZ PRESS

DARK DORSET CALENDAR CUSTOMS
Robert J Newland - ISBN 978-1-905723-18-8

£12.50

Much of the intrinsic charm of Dorset folklore is owed to the importance of folk customs. Today only a small amount of these curious and occasionally eccentric customs have survived, while those that still continue have, for many of us, lost their original significance. Why do we eat pancakes on Shrove Tuesday? Why do children dance around the maypole on May Day? Why do we carve pumpkin lanterns at Hallowe'en? All the answers are here! Robert has made an in-depth study of the Dorset country calendar identifying the major feast-days, holidays and celebrations when traditionally such folk customs are practiced.

CENTRE FOR FORTEAN ZOOLOGY 2004 YEARBOOK
Edited by Jonathan Downes and Richard Freeman
ISBN 978-1-905723-14-0

£12.50

The Centre For Fortean Zoology Yearbook is a collection of papers and essays too long and detailed for publication in the CFZ Journal *Animals & Men*. With contributions from both well-known researchers, and relative newcomers to the field, the Yearbook provides a forum where new theories can be expounded, and work on little-known cryptids discussed.

CENTRE FOR FORTEAN ZOOLOGY 2008 YEARBOOK
Edited by Jonathan Downes and Corinna Downes
ISBN 978-1-905723-19-5

£12.50

The Centre For Fortean Zoology Yearbook is a collection of papers and essays too long and detailed for publication in the CFZ Journal *Animals & Men*. With contributions from both well-known researchers, and relative newcomers to the field, the Yearbook provides a forum where new theories can be expounded, and work on little-known cryptids discussed.

ETHNA'S JOURNAL
Corinna Newton Downes
ISBN 978-1-905723-21-8

£9.99

Ethna's Journal tells the story of a few months in an alternate Dark Ages, seen through the eyes of Ethna, daughter of Lord Edric. She is an unsophisticated girl from the fortress town of Cragnuth, somewhere in the north of England, who reluctantly gets embroiled in a web of treachery, sorcery and bloody war...

**CFZ PRESS, MYRTLE COTTAGE,
WOOLFARDISWORTHY BIDEFORD,
NORTH DEVON, EX39 5QR
www.cfz.org.uk**

Other books available from
CFZ PRESS

ANIMALS & MEN - Issues 11 - 15 - The Call of the Wild
Jonathan Downes (Ed) - ISBN 978-1-905723-07-2

£12.50

Since 1994 we have been publishing the world's only dedicated cryptozoology magazine, *Animals & Men*. This volume contains fascimile reprints of issues 11 to 15 and includes articles covering out of place walruses, feathered dinosaurs, possible North American ground sloth survival, the theory of initial bipedalism, mystery whales, mitten crabs in Britain, Barbary lions, out of place animals in Germany, mystery pangolins, the barking beast of Bath, Yorkshire ABCs, Molly the singing oyster, singing mice, the dragons of Yorkshire, singing mice, the bigfoot murders, waspman, British beavers, the migo, Nessie, the weird warbling whatsit of the westcountry, the quagga project and much more...

IN THE WAKE OF BERNARD HEUVELMANS
Michael A Woodley - ISBN 978-1-905723-20-1

£9.99

Everyone is familiar with the nautical maps from the middle ages that were liberally festooned with images of exotic and monstrous animals, but the truth of the matter is that the *idea* of the sea monster is probably as old as humankind itself.

For two hundred years, scientists have been producing speculative classifications of sea serpents, attempting to place them within a zoological framework. This book looks at these successive classification models, and using a new formula produces a sea serpent classification for the 21st Century.

CENTRE FOR FORTEAN ZOOLOGY 1999 YEARBOOK
Edited by Jonathan Downes
ISBN 978 -1-905723-24-9

£12.50

The Centre For Fortean Zoology Yearbook is a collection of papers and essays too long and detailed for publication in the CFZ Journal *Animals & Men*. With contributions from both well-known researchers, and relative newcomers to the field, the Yearbook provides a forum where new theories can be expounded, and work on little-known cryptids discussed.

CENTRE FOR FORTEAN ZOOLOGY 1996 YEARBOOK
Edited by Jonathan Downes
ISBN 978 -1-905723-22-5

£12.50

The Centre For Fortean Zoology Yearbook is a collection of papers and essays too long and detailed for publication in the CFZ Journal *Animals & Men*. With contributions from both well-known researchers, and relative newcomers to the field, the Yearbook provides a forum where new theories can be expounded, and work on little-known cryptids discussed.

CFZ PRESS, MYRTLE COTTAGE, WOOLFARDISWORTHY BIDEFORD, NORTH DEVON, EX39 5QR
w w w . c f z . o r g . u k

Other books available from
CFZ PRESS

BIG CATS IN BRITAIN YEARBOOK 2008
Edited by Mark Fraser - ISBN 978-1-905723-23-2

£12.50

People from all walks of life encounter mysterious felids on a daily basis, in every nook and cranny of the UK. Most are jet-black, some are white, some are brown; big cats of every description and colour are seen by some unsuspecting person while on his or her daily business. 'Big Cats in Britain' are the largest and most active research group in the British Isles and Ireland. This book contains a run-down of every known big cat sighting in the UK during 2007, together with essays by various luminaries of the British big cat research community.

CFZ EXPEDITION REPORT 2007 - GUYANA
ISBN 978-1-905723-25-6

£12.50

Since 1992, the CFZ has carried out an unparalleled programme of research and investigation all over the world. In November 2007, a five-person team - Richard Freeman, Chris Clarke, Paul Rose, Lisa Dowley and Jon Hare went to Guyana, South America. They went in search of giant anacondas, the bigfoot-like didi, and the terrifying water tiger.

Here, for the first time, is their story...With an introduction by Jonathan Downes and forward by Dr. Karl Shuker.

CENTRE FOR FORTEAN ZOOLOGY 2003 YEARBOOK
Edited by Jonathan Downes and Richard Freeman
ISBN 978 -1-905723-19-5

£12.50

The Centre For Fortean Zoology Yearbook is a collection of papers and essays too long and detailed for publication in the CFZ Journal *Animals & Men*. With contributions from both well-known researchers, and relative newcomers to the field, the Yearbook provides a forum where new theories can be expounded, and work on little-known cryptids discussed.

CENTRE FOR FORTEAN ZOOLOGY 1997 YEARBOOK
Edited by Jonathan Downes and Graham Inglis
ISBN 978 -1-905723-27-0

£12.50

The Centre For Fortean Zoology Yearbook is a collection of papers and essays too long and detailed for publication in the CFZ Journal *Animals & Men*. With contributions from both well-known researchers, and relative newcomers to the field, the Yearbook provides a forum where new theories can be expounded, and work on little-known cryptids discussed.

**CFZ PRESS, MYRTLE COTTAGE,
WOOLFARDISWORTHY BIDEFORD,
NORTH DEVON, EX39 5QR
www.cfz.org.uk**

Other books available from
CFZ PRESS

CENTRE FOR FORTEAN ZOOLOGY 2000-1 YEARBOOK
Edited by Jonathan Downes and Richard Freeman
ISBN 978-1-905723-19-5

£12.50

Edited by
Jonathan Downes and Richard Freeman

The Centre For Fortean Zoology Yearbook is a collection of papers and essays too long and detailed for publication in the CFZ Journal *Animals & Men*. With contributions from both well-known researchers, and relative newcomers to the field, the Yearbook provides a forum where new theories can be expounded, and work on little-known cryptids discussed.

CENTRE FOR FORTEAN ZOOLOGY 2002 YEARBOOK
Edited by Jonathan Downes and Richard Freeman
ISBN 978-1-905723-30-0

£12.50

Edited by
Jonathan Downes
and
Richard Freeman

The Centre For Fortean Zoology Yearbook is a collection of papers and essays too long and detailed for publication in the CFZ Journal *Animals & Men*. With contributions from both well-known researchers, and relative newcomers to the field, the Yearbook provides a forum where new theories can be expounded, and work on little-known cryptids discussed.

CFZ PRESS, MYRTLE COTTAGE,
WOOLFARDISWORTHY BIDEFORD,
NORTH DEVON, EX39 5QR
www.cfz.org.uk

www.ingramcontent.com/pod-product-compliance
Lightning Source LLC
Chambersburg PA
CBHW060656100426
42734CB00047B/1926